Extrusion cooking

Related titles from Woodhead's food science, technology and nutrition list:

Advances in cereals processing (ISBN: 1 85573 561 X)

This book is a major new review of key technological advances in cereals processing, covering the range of cereals products from bread making, pasta and rice production to breakfast cereals and starch products. This will be essential reading for raw material producers and primary and secondary cereal processors.

Thermal technologies in food processing (ISBN: 1 85573 558 X)

Thermal technologies are traditionally a compromise between their benefits in the enhancement of sensory characteristics and preservation, and their shortcomings, for example in reducing nutritional properties. The need to maximise process efficiency and final product quality has led to a number of new developments including refinements in existing technologies and the emergence of new thermal techniques. This collection reviews all these key developments and looks at future trends, providing an invaluable resource for all food processors.

Food processing technology: Principles and practice (ISBN: 1 85573 533 4)

The first edition of *Food processing technology* was quickly adopted as the standard text by many food science and technology courses. The publication of this completely revised new edition is set to confirm the position of this textbook as the best single-volume introduction to food manufacturing technologies available. New chapters include computer control of processing, novel 'minimal' technologies including processing using high pressures or pulsed electric fields, ohmic heating and an extended chapter on modified atmosphere packaging.

Extrusion cooking

Technologies and applications

Edited by
Robin Guy

CRC Press
Boca Raton Boston New York Washington, DC

WOODHEAD PUBLISHING LIMITED
Cambridge England

Published by Woodhead Publishing Limited, Abington Hall, Abington
Cambridge CB21 6AH, England
www.woodheadpublishing.com

Published in North America by CRC Press LLC, 6000 Broken Sound Parkway, NW,
Suite 300, Boca Raton, FL 33487, USA

First published 2001, Woodhead Publishing Limited and CRC Press LLC
Reprinted 2007

British Library Cataloguing in Publication Data
A catalogue record for this book is available from the British Library.

Library of Congress Cataloging in Publication Data
A catalog record for this book is available from the Library of Congress.

Woodhead Publishing ISBN 978-1-85573-559-0 (book)
Woodhead Publishing ISBN 978-1-85573-631-3 (e-book)
CRC Press ISBN 978-0-8493-1207-6
CRC Press order number: WP1207

The publishers' policy is to use permanent paper from mills that operate a sustainable forestry policy, and which has been manufactured from pulp which is processed using acid-free and elementary chlorine-free practices. Furthermore, the publishers ensure that the text paper and cover board used have met acceptable environmental accreditation standards.

Printed by TJI Digital, Padstow, Cornwall, England

Contents

Contributors

Chapter 1, 2 and 8

Dr Robin Guy
Campden and Chorleywood Food Research Association
Chipping Campden
Glos GL55 6LD

Tel: +44 (0)1386 842136
Fax: +44 (0)1386 842100
E-mail: *r.guy@campden.co.uk*

Chapter 3

Professor Mian N. Riaz
Head – Extrusion Technology Program
Food Protein R & D Center
Texas A&M University
College Station
TX 77843-2476
USA

Tel: (1) 979 845 2774
Fax: (1) 979 845 2744
E-mail: *mnriaz@tamu.edu*

Chapter 4

Dr Jérôme Mottaz
Clextral
B.P.10 – ZI de Chazeau
F-42702 Firminy Cedex
France

Tel: +33 (0) 4 77 40 31 31
Fax: +33 (0) 4 77 40 31 00
E-mail: *clextralig@aol.com*

Chapter 5

Dr Jay Sellahewa
Food Engineering Group
Food Science Australia
PO Box 52
North Ryde
NSW 1670
Australia

Fax: +61(0) 2 9490 8310
E-mail: *Jay.Sellahewa@foodscience.afisc.csiro.au*

Dr Charlie Chessari
Principal Engineer
Advanced Process Control
Foxboro Australia
PO Box 155
Alexandria NSW 1435
Australia

Tel: +61 (0) 2 8396 3660
Fax: +61 (0) 2 9690 1845
Mobile: +61 (0) 419 596 384
E-mail: *cchessar@foxboro.com.au*

Chapter 6

Professor Mary Ellen Camire
Department of Food Science and Human Nutrition
University of Maine
5736 Holmes Hall
Orono
ME 04469-5736
USA

Tel: (1) 207 581 1627
Fax: (1) 207 581 1636
E-mail:
MaryCamire@umit.maine.edu

Chapter 7

Professor Jean-Marie Bouvier
Clextral
B.P.10 – ZI de Chazeau
F-42702 Firminy Cedex
France

Tel: +33 (0) 4 77 40 31 31
Fax: +33 (0) 4 77 40 31 00
E-mail: *jmbouvier@clextral.com*

Chapter 9

Dr Massoud Kazemzadeh
Buhler Inc
10025 Beard Ave South
Bloomington
MN 55431
USA

Tel: (1) 952 831 0908
Fax: (1) 952 831 1211

1

Introduction

R. Guy, Campden and Chorleywood Food Research Association, Chipping Campden

Extrusion technologies have an important role in the food industry as efficient manufacturing processes. Their main role was developed for conveying and shaping fluid forms of processed raw materials, such as doughs and pastes. Extrusion cooking technologies are used for cereal and protein processing in the food and, closely related, petfoods and feeds sectors. The processing units have evolved from simple conveying devices to become very sophisticated in the last decade. Today, their processing functions may include conveying, mixing, shearing, separation, heating or cooling, shaping, co-extrusion, venting volatiles and moisture, flavour generation, encapsulation and sterilisation. They can be used for processing at relatively low temperatures, as with pasta and half-product pellet doughs, or at very high ones with flatbreads and extruded snacks. The pressures used in extruders to control shaping, to keep water in a superheated liquid state and to increase shearing forces in certain screw types, may vary from around 15 to over 200 atmospheres.

The most important feature of an extrusion process is its continuous nature. It operates in a dynamic steady state equilibrium, where the input variables are balanced with the outputs. Therefore, in order to obtain the required characteristics in an extrudate, the multivariate inputs must be set at the correct levels to give the dependent physical conditions and chemical process changes within the barrel of the machine. These dependent system variables determine the extrudate variables, which are reflected in the product variables. Once the relationships between the independent variables and the dependent variables within the processor are established for an individual product type, they must be maintained close to their optimum levels, in a small processing window, to ensure that the extrudate variables are also kept at the required levels.

In the development of extrusion processes, there have been improvements in extrusion equipment and the ancillary processing units, which form a complete processing line for an individual product type. However, in recent years the development of a better understanding of the sequence of processes occurring within extruders has been equally important. For example, the establishment of the role of individual machine variables on commercial extruders and the measurement of the effects of raw materials on dependent processing variables. Raw materials were shown to play an active role in determining the magnitude of variables such as pressure, temperature and motor load, as well as providing the structure forming materials, which are developed in the extruder to form the extrudate. This new understanding within the industry has led to better use of existing machines and modification to improve their function. The development of extruders has moved forward from a purely empirical approach, which led to the development of products on the early single screw machines from 1940 onwards. Extrusion technologists are now more likely to use mathematical modelling on different applications of extruders.

Extrusion cooking has gained in popularity over the last two decades for a number of reasons:

- versatility: a wide range of products, many of which cannot be produced easily by any other process, is possible by changing the ingredients, extruder operating conditions and dies
- cost: extrusion has lower processing costs and higher productivity than other cooking and forming processes
- productivity: extruders can operate continuously with high throughput
- product quality: extrusion cooking involves high temperatures applied for a short time, retaining many heat sensitive components of a food
- environmentally-friendly: as a low-moisture process, extrusion cooking does not produce significant process effluents, reducing water treatment costs and levels of environmental pollution.

Therefore, this book looks at the range of important variables which influence the processing window during manufacturing and which must be maintained within fairly narrow limits to produce high quality products, with respect to sensory and nutritional characteristics. These process variables will include both machine variables and raw material characteristics. In their optimisation to create a product can lie the success or failure of an extruded product. Examples are given for a number of important technologies from breakfast cereals, snackfoods to baby foods.

Part I

General influences on quality

2

Raw materials for extrusion cooking

R. Guy, Campden and Chorleywood Food Research Association, Chipping Campden

2.1 Introduction

2.1.1 General nature of raw materials used in extrusion

Extruded foods and feeds are made from a wide and diverse range of raw materials. These ingredients are similar in their general nature to the ingredients used in all other types of foods and feeds. They contain materials with different functional roles in the formation and stabilisation of the extruded products, and provide colour, flavours and nutritional qualities found in different product types. The transformation of raw materials during processing is one of the most important factors that distinguishes one food process and food type from another. For a particular product type a selection of ingredients is processed through a set processing regime. For extrusion cooking this involves heating to high temperatures, the application of mechanical mixing and shearing, before finally extruding to form a structure. If conditions are in the ideal processing range a stable extrudate will form with the normal product characteristics required for that product.

Extrusion cooking is a specialised form of processing,[1] which is unique in food and feed processing because of the conditions that are used to transform the raw materials. It is a relatively low moisture process compared with conventional baking or dough processing. Normal moisture levels used are in the range of 10–40% on a wet weight basis. Despite these low moistures the mass of raw materials is transformed into a fluid and subjected to a number of operations to mix and transform the native ingredients into new functional forms. Under these unusual process conditions the physical features of raw materials, such as the particle size, hardness and frictional characteristics of powders and the lubricity and plasticising power of fluids become more important than in other food and feed processes (see Fig. 2.1).

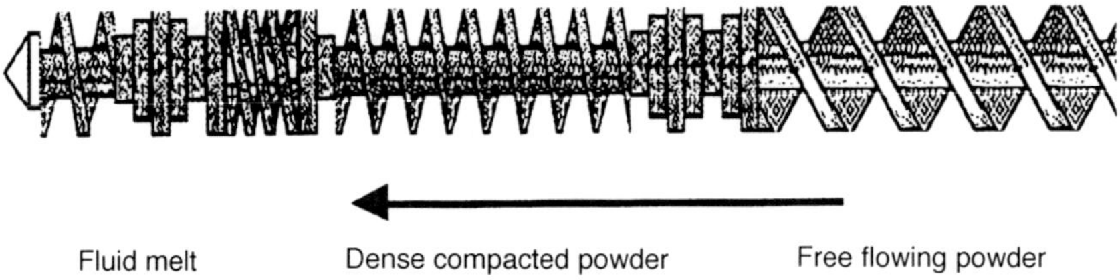

Fig. 2.1 Changes in raw materials in an extrusion cooking process.

A second feature that distinguishes extrusion cooking from other food processes is the use of very high temperatures, usually in the range 100–180ºC. The aqueous dough systems are superheated and the water vapour is contained within the extruder at high pressure. The use of high temperatures reduces the processing time and allows a full transformation of raw material to its functional form in periods as little as 30–120 s. Almost all extrusion cooking processes are operated continuously with raw materials fed into the processing units. The products may be created by extrusion from dies to form the required product structure in direct extrusion, or to form the half-products in the second generation snack pellets.

All food and feed products have basic structures that are formed by certain elements in the raw materials such as the biopolymers of starch and proteins in baked products, or fat and sugar in confectionery. The structural elements form the three-dimensional cages or nest of girders in which the other materials are held to form the product texture. Extruded products are formed from the natural biopolymers of raw materials such as cereal or tuber flours[2] that are rich in starch, or oilseed legumes and other protein-rich sources. The most commonly used materials are wheat and maize flours, but many other materials are also used such as rice flour, potato, rye, barley, oats, sorghum, cassava, tapioca, buckwheat, pea flour and other related materials.

If the extruded products are manufactured in the form of texturised vegetable protein (TVP) the main ingredients will be selected from protein-rich materials such as pressed oilseed cake from soya, sunflower, rape, field bean, fava beans, or separated proteins from cereals such as wheat (gluten).

The native forms of the biopolymers were not designed for extrusion cooking and must be changed by processing to obtain a more useful polymer size and form for structure creation as a desirable product. All the natural biopolymers in the ingredients listed above can be transformed into a fluid melt in the temperature and moisture ranges used in an extruder. The skill in controlling the processing is to transform the polymers in a short period of time using the thermomechanical processing provided by the screw elements under the control of the die pressure. In a normal recipe all the ingredients will interact with one another to affect the transformations taking place. Therefore, it is important to understand the role of each individual material in the recipe and the effect of any variation in an individual ingredient on the overall processing performance of the extruder.

2.1.2 Classification of ingredients by their functional roles in extrusion cooking

The complex mixture of materials present in a recipe may appear very confusing to the extrusion cooking technologist and machine operator. One of the first steps taken at CCFRA in developing a better understanding of the extrusion cooking process was the introduction of the Guy Classification System for ingredients. This was published in 1994[3] and is based on the grouping of ingredients according to their functional role using a physicochemical approach. Originally six groups were selected to describe the functional roles of all the ingredients but one group has been subdivided to increase the number to seven.

Group 1: Structure-forming materials

The structure of an extruded product is created by forming a melt fluid from biopolymers and blowing bubbles of water vapour into the fluid to form a foam. The film of biopolymers must flow easily in the bubble walls to allow the bubbles to expand as the superheated water is released very quickly at atmospheric pressure. Fluid melts of biopolymers form the cell walls of gas bubbles and allow them to extend until they burst. After expansion, the rapid fall in temperature caused by evaporation, and the rise in viscosity due to moisture loss, rigidifies the cellular structure. The rapid increase in viscosity is followed by the formation of a glassy state. Starch polymers are very good at this function and well-expanded cellular structures can be made from any of the separated starches available from materials such as wheat, maize, rice or potato. The average polymer size found in most natural starches is far too large for the optimum expansion. The most abundant polymer amylopectin has a molecular weight of up to 10^8D, which gives poor flow properties in gas cell walls and low expansion (1–2 ml/g). However, the use of high levels of mechanical shear during extrusion cooking can reduce the average molecular weight of AP to $< 10^6$D. The smaller molecules allow much more flow in bubble cells walls and cause an increase in expansion from 1 to 25 ml/g. The natural starch from amylomaize, which contains a large proportion of the smaller starch polymer amylose (2–10^5D), gives the largest expansion of the native starches.

Structure forming polymers must have a minimum molecular weight sufficient to give enough fluid viscosity to prevent or control the shrinkage of an extrudate after it has reached its maximum expansion and ruptured the gas cells. If the extrudate is too viscous at this point there will be rapid shrinkage and loss of apparent expansion in extrudates. This occurs when starch polymers are reduced in size to form maltodextrins of dextrose equivalent, DE 10 to 20. At this stage their viscosity is too low at the moisture levels used in extrusion either to induce rupture or stabilise the cell walls against elastic recoil effects. Their extrudates will collapse after expansion due to low internal pressure in the unbroken bubbles or low viscosity to give little apparent expansion on cooling. Therefore, they are not classified as structure-forming materials.

Proteins may be used to form structures in extrudates at high concentrations. For example soya proteins may be used to produce an expanded structure in TVP, if their concentration in the recipe is > 40% w/w, at moisture levels of 30–

40% w/w. They are globular proteins of 80–100 × 10^3D, significantly smaller than starch polymers in the melt fluid but they may be linked together to form larger structures as they flow through a die channel. They aggregate and form higher viscosity complexes, which serve to form crude films and retain some of the expanding water vapour. Their viscosity on cooling is sufficient to prevent shrinkage and allows an alveolar structure to be formed. Other proteins, which also undergo similar transitions, are found in legumes and in the endosperm of wheat flour. Wheat gluten is a hydrophobic protein, which can form polymers of greater molecular weight than the native form.

Group 2: Dispersed-phase filling materials

The examination of microscopic sections through extruded products, such as snacks or petfoods made from starch-rich recipes, shows a continuous phase of starch polymers. However, several dispersed phases lie within the continuous starch structure. The most obvious of these will be formed by any proteins present and by fibrous materials such as cellulose or bran. The proteins may be present in several forms depending on the ingredients used, as they may be derived from cereals, legumes or animal proteins. These polymers will form separate phases within the continuous starch phase. Their size and shape in a particular product will depend on their original particle size and their resistance to shear during processing.

Proteins such as gluten (added at levels $< 30\%$), which hydrate in water and become soft doughs, will be reduced in size by the screws in proportion to the severity of the processing and may be as small as 5 μm after processing. Water-soluble proteins such as albumins will coagulate at high temperature and then the coagulum will be broken down in a similar manner to a similar size range.

Fibrous materials found in an extrusion cooking recipe would include materials comprised of hemi-cellulose, cellulose and lignin derived from the husks and bran of grains and seeds. These materials tend to remain firm and stable during processing and are not reduced in size during extrusion.

In all cases, the presence of the dispersed-phase materials affects the nature of the extrusion process in two ways. Their physical presence in the cell walls will reduce the potential for expansion of the starch film by disrupting the cell walls when their structures penetrate the walls of the film. This effect is easily observed with wheat bran, which may have an average particle size of 0.8–2 mm. The bran particles have little effect at low concentration of 1 to 2%, but when added at levels found in a wholemeal flour (8 to 9% bran), may reduce both expansion and apparent expansion by as much as 50%. At higher levels of added bran the expansion falls with concentration until it disappears at about 75% added bran, when there is insufficient starch present to form a continuous film to contain the water vapour.

The second effect caused by the presence of dispersed filler relates to the elastic recoil or die swell effect of the fluid as it leaves the die exit. Pure starch fluids are very elastic and when they are deformed as they enter the die, they store the elastic energy in their molecular structures. This energy is released as the fluid leaves the die and causes a swelling effect normal to the direction of flow in the die. It has been observed in plastics research that the presence of inert fillers, such as carbon black, reduces the die swell in plastics extrusions until it disappears at concentrations of 30–40% added filler. A similar effect was also detected in recipes containing added proteins or bran in wheat starch extrusion so that at similar filler levels the die swell effect was eliminated.

Group 3: Ingredients that act as plasticisers and lubricants

In low moisture doughs used for extrusion cooking the initial physical interactions in the recipe cause frictional and mechanical energy dissipation. This energy source serves to heat the dough mass. The heating rate is very high in low moisture systems, so that for recipes up to 25% moisture no external heating is required to reach an operating temperature of 150°C. The addition of ingredients such as water serves to reduce interactions by plasticising the dry polymer forms, transforming them from solids to deformable plastic fluids. The addition of increasing amounts of water reduces the dissipation of mechanical energy and reduces the heat input as the moisture level is increased.

The particles of starch, fibre and proteins are mechanically sheared by the screws system of the extruder to change their physical form. The levels of applied shear may be reduced by the presence of oils and fats. These materials serve to lubricate both the interacting particles in the dough mass and the particles that are rubbing against the metal surfaces of screws and barrel. The effect of lubricants is more powerful than that of plasticisers in terms of their active concentrations.

Oils and fats produce large effects on the processing of starch at levels of 1–2% and higher levels may reduce the degradation of the starch polymer to such an extent that no expansion is obtained from a recipe. In certain recipes the effect of high levels of oils and fats is reduced by the addition of materials that can absorb the lipids in hollow rigid structures such as bone meal.

Group 4: Soluble solids

Some low molecular weight materials, such as sugars or salts, may be added to a recipe for flavouring or humectant properties. The materials that are soluble will dissolve in free dough water during the initial mixing stage of processing. Their effect on the extrusion process will depend on their concentration and their chemical interaction with starch and protein polymers. All small molecules added to a recipe must dilute other ingredients. If they replace starch the viscous effect of the large polymers will be reduced and the hot melt fluid will become less viscous unless the water levels are reduced. In direct action on the polymers only strong acids have been shown to have a significant effect on the degradation of starch.

Group 5: Nucleating substances

Substances that increase bubble nucleation have been found to increase the numbers of bubbles appearing in the hot melt fluid of an extruder. Two well known materials have been demonstrated to increase the numbers of bubbles present in an expanding extrudate, powdered calcium carbonate 'creta preparata' and talc (magnesium silicate). The addition of a finely powdered material that remains insoluble in the dough provides surfaces to reduce the energy required for the formation of individual bubbles and can increase their numbers from a few hundred to over 70×10^3 per ml.

Group 6: Colouring substances

Materials may be added to the recipe to produce colour in the extrudates. These would include heat stable colours and the precursors of colour formation by thermal reactions. Some colour may be found in natural raw materials such as maize, which can be used to add to the palette.

Group 7: Flavouring substances

The flavouring of extruded products follows a similar pattern to colouring. Flavour compounds may be added either during extrusion or in secondary operations post extrusion. In addition precursors to flavour formation in thermal reactions may be added to form flavours in the extruder.

Tables 2.1 and 2.2 give examples of recipes for extruded products from snackfoods and breakfast cereals.

Table 2.1 Snackfoods

Group	Common name	Corn curl	Maize/potato	Wheat
1	Maize grits	80.7	50	–
	Potato granules	–	20	–
	Potato starch	–	5	–
	Wheat flour	–	–	70
2	Wheat gluten	–	2	–
	Soya flour	–	–	5
	Wheat bran	–	–	10
3	Vegetable oil	0.5	1.5	1
	Water	16	18	16
	Monoglyceride	0.3	0.3	0.3
4	Salt	1	1	1.5
	Maltodextrins		5	
5	Calcium carbonate	–	1.5	–
6	Milk powder	1	2	2.5
7	Glucose/ peptides	0.15	0.15	0.1

Table 2.2 Breakfast cereals

Group	Common name	Corn flake	Crisp rice	Multigrain
1	Maize grits	80.7	50	–
	Rice flour	–	20	–
	Oat flour	–	5	–
	Wheat flour	–	–	70
2	Wheat bran	–	–	10
3	Vegetable oil	0.5	1.5	1
	Water	16	18	16
	Monoglyceride	0.3	0.3	0.3
4	Salt	1	1	1.5
	Sugar	5	5	5
5	Calcium carbonate	–	1.5	–
6	Milk powder	1	2	2.5
7	Malt	0.15	0.15	0.1

2.2 Examples from Group 1: structure-forming materials based on starch

The structure of extruded products may be formed from either starch or protein biopolymers. The majority of products in the breakfast cereal, snackfood and biscuit markets are formed from starch. Proteins are only used to form products that have meat-like characteristics and are used either as full, or partial, replacements for meat in ready meals, dried foods and many petfood products.

2.2.1 Starch-based products

The main sources of starch are from the abundant cereal and potato crop of the major industrialised countries of the world. There may be some other crops such as cassava or sago that are also a good source of starch in their countries of origin, but for most manufacturers the large cereal crops are the most economic source of starch. Potato derivatives have unique flavour qualities, which add value to the products in which they are to cover their higher costs. Examples of starch-based products are given in Table 2.3.

Factors which affect the performance of starch in extrusion cooking

Research studies at CCFRA since 1983 has been aimed at finding out which are the most important factors in starch. In order to do so we have followed the starch granules through the extrusion cooking processes using a twin screw extruder equipped with monitoring probes for temperature and pressure and a clam shell barrel to examine the development of material along a standard screw configuration. The main extrusion process for a starch-rich recipe was found to follow a standard sequence irrespective of a particular material or recipe. This was as follows:

1. The powders are mixed with water and conveyed to the compression zone.
2. The powder is compressed to a density of 1 g/ml at 5–10 bar.
3. The powder is heated by frictional and mechanical dissipation of energy from the motor and heat from the barrel.
4. The starch granules melt and become soft.
5. The starch granules are compressed to a flattened form.
6. The starch polymers are dispersed and degraded to form a continuum in the melt fluid.
7. The starch polymer continuum holds and stretches with the expanding bubbles of water vapour during extrudate expansion until the rupture point is reached.
8. The starch polymer cell walls recoil and stiffen as they cool to stabilise the extrudate structure.
9. The starch polymers become glassy as the moisture is removed to form a hard brittle texture.

Thus the essential features of the process are the melting of the crystalline regions of granules and the release and degradation of the starch polymers. The

Table 2.3 Examples of starch-based products

Class	Product type	Starch sources
Breakfast cereals	Corn flake	maize
	Multigrain flake	wheat, maize, rice, oats
	High fibre flake	maize, wheat, rice
	Crisp rice	rice
	Oat puffs	oats
Snacks	Potato sticks	potato, maize
	Potato hoops	potato
	Corn puffs or curls	maize
	Puffs	wheat, potato, maize
	Prawn cracker	wheat, rice
	Half-products	maize, wheat, potato, rice
	Mexican corn chips	maize
	Mexican style chips	wheat, potato
Biscuits	Flatbread	wheat, maize, rice cassava
Petfoods	Dry cat and dog shapes	wheat, maize, rice
	Moist morsels	wheat
	Dry treats	wheat, maize
Fish feeds	Floating and sinking for fish	wheat
	Sunken types for prawns	wheat

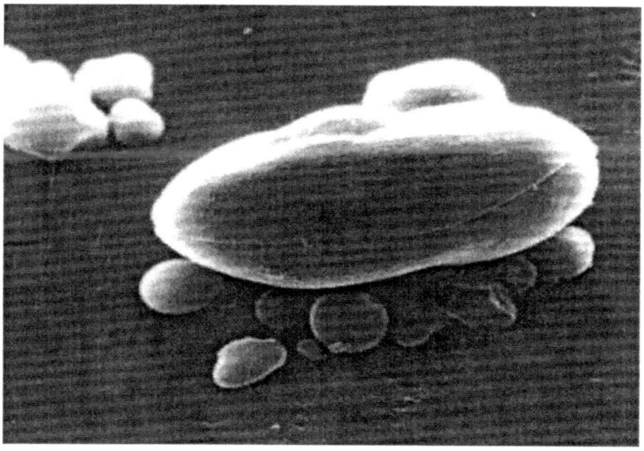

Fig. 2.2 Scanning electron micrograph of wheat starch granules showing large lenticular and small globular granules (from studies by A. E. Evers of CCFRA).

differences in processing are caused by differences in the melting characteristics due to size and shape and by the polymers which are dispersed from the granules. In order to understand this more precisely we must examine the nature of starch granules.

Physical form and composition of natural starches from cereals and tubers

The common physical form of starch in all cereals and tubers is small aggregates of polymer molecules known as granules. However, the size and shape of starch granules vary widely from plant to plant (Fig. 2.2). In rice kernels, small globular granules of 1–10 μm are found in clusters, whereas maize kernels contain both globular polygonal granules of 15–20 μm. Potatoes have large flattened globular granules of 50–100 μm. A series of microgaphs of starch granules published in the literature[4,5] shows that all common plants have differences in the morphological form of their granules.

A special feature of all starch granules is that portions of the polymers are present as crystalline regions. These regions are present in structures of granules serving to hold the polymer mass together in a rigid form, so that the granules do not swell in water and behave as hard bodies forming a rheopectic paste. The crystalline regions also serve to rotate the plane of polarised light introducing birefringence (the Maltese Cross effect), when the starch granules are observed between cross-polarised filters on a light microscope.

The physical nature of the starch granules

The physical size and shape of starch granules may affect certain extrusion processes. Small starch granules have shorter distances for the heat to travel to raise their temperature to the critical melting point and therefore become soft more quickly in the extruder barrel. This effect is observed most clearly when comparing the melting of rice starch with wheat or maize starches. In low moisture conditions 14–16% w/w, a second effect may be observed due to the physical shape of starch granules. If wheat starch is compared with maize at low moisture and low screw speed the energy inputs are fairly similar, but at high screw speed the maize creates almost twice as much specific mechanical energy (SME) as the wheat starch. This is probably because of the polygonal shape of almost half of the starch granules found in normal samples of maize compared with the smooth globular or lenticular granules of wheat.

Composition of granules

The major constituents of starch granules, representing 97–98% of the dry matter, are the two physical forms of starch, amylose (AM) and amylopectin (AP). In addition there are trace amounts of lipid (0.5–1%), mainly in the form of lysolecithin, dispersed throughout the starch rings and 0.1 to 0.2% of proteins adhering to the surface layers.[5] The two starch polymers each represent a polydisperse family in terms of the molecular size range.

Amylose (AM) polymers range in size from 100 to 200 kD and are linear polymers with one or two branches at C-6 of a glucose residue. They complex with lysolecithin and similar monoacyl lipids of saturated fatty acids > 6 carbon atoms. There are variations in AM levels in starches from different cereals and tubers and within varieties from the same types. In wheat, rye and oats the levels of AM are normally within 20–27% of the total starch but in maize, barley and rice it can vary from 5 to 30% and for maize it can be as high as 70%. The highest levels are found in the amylomaize and the lowest forms in the waxy barley, maize or rice.

Starch polymers are formed in nature from glucose units, during the development of the seeds on the plants. A linear polymer of 100 to 200 glucose units is formed which has been called amylose (AM). Some of the linear chains are built up into larger branched chain molecules called amylopectin (AP). These are very large polydisperse molecules up to 10^8D in molecular weight containing 500,000 glucose units. The structure of AP is dendritic with a main chain resembling the trunk of a tree and secondary chains as the larger branches and many smaller chains as the fine twig-like branches. These outer chains are 20 glucose units in length and can link the AP trees together in intermolecular double helices to form a rigid structure.

Amylopectin (AP) polymers are much larger than AM and their molecular ranges are difficult to measure because the procedures used to isolate these giant molecules can easily lead to degradation. The ranges reported are from 2 to 10^8D but there are no accurate comparisons of AP from different cereal or tubers types. The interactions between the chains are difficult to release without breaking the main chains and reducing the polymer size. It is easier to measure the lengths of the side chains by hydrolysing the whole molecule with a debranching enzyme such as pullulanase. This enzyme releases all the starch chains from the main tree-like structure as small molecules of 16–20 glucose units or 3 to 4 kD. These relatively short outer chains serve to form the intermolecular bonds between individual AP molecules and are used to explain the differences in stability to retrogradation of different forms of cereal starch.

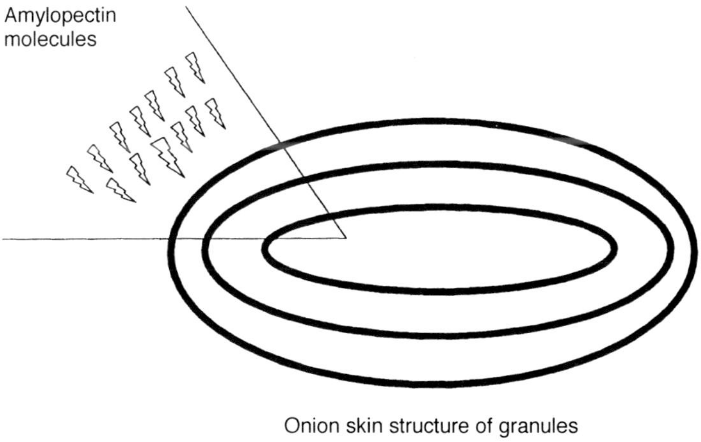

Fig. 2.3 Diagram of a starch granule showing the layered structure and arrangement of amylopectin molecules.

One other feature of AP molecules that differentiates tuber starches such as potato or cassava from cereal starches is the presence of phosphate groups on the AP from tuber starches. These ionic groups serve to increase the water affinity of the starch giving a larger swelling power for these starches in water. However, the water affinity is also strongly influenced by salt and acids, which affect the ionisation of the phosphate groups.

The general composition of starch in common cereals is a ratio of 3 to 1 for AP to AM. The AP is laid down in rings with the reducing end on the inside, to form an onion-like structure held together by the linkages in the peripheral linkages in rings (Fig. 2.3). The wheat starch system creates small spherical bodies in the endosperm cells of the plant seed, which are known to all cereal microscopists as granules. The granules grow to about 10 μm in diameter and then some may fuse together to form a larger lenticular granule of 20–40 μm in maximum dimension (Fig. 2.2).

Sources of starch: the cereals

The three major cereals in order of world production are wheat, rice and maize but there are also substantial crops of barley, rye, triticale, oats and sorghum. All these cereals are available as grains, which may be milled to form fractions of the inner endosperm and the outer layer of hull or pericarp. The most common form of the cereals used in human foods are flours which have had a large proportion of the hulls removed. This procedure enriches the starch fraction and removes fibrous material hemicellulose and beta glucan together with some polyphenolic materials such as lignin.

The composition of the materials that can be purchased from the miller is shown in Table 2.4 compared with wholemeal flour prepared by hammer milling whole grain. The composition of the raw material varies with the level of

Table 2.4 Composition of materials prepared from wheat grain as per cent of dry solids*

Component	Whole grain	Semolina	Flour	Starch
Protein	12	11	11	0.2
Starch	75	78	80	90
Fibre	8.0	2.0	1.5	0.5
Lipid	2.5	1.5	1.5	1.0
Ash	1.5	1.0	1.0	0.5

* Data from samples examined at CCFRA,

refinement from grain to washed starch. In the grain the starch granules are found in the endosperm region where it is formed in the plant cells together with the cereal proteins. As the endosperm is separated from the outer layer of the grain during milling the amount of structure-forming material increases with the purification process. The most refined product, washed starch, still contains some impurities but these are buried with the structure of the granular bodies in which the starch is laid down in the cereals (see Figs 2.4–2.6).

The cost of the starch will increase as it is processed; therefore it is more economic to use starch sources that have the other materials present, if possible. The other materials shown in Table 2.4 can classified as fillers (protein and fibres) or lubricants (lipids) and should be considered in these roles in relation to the rest of the recipe being used.

The composition of different types of cereals varies with the levels of non-starch components (see Table 2.5). Maize and rice flours are generally richer in starch than wheat flour due to the lower protein and fibre levels. Oat flours prepared from the rolled flakes are high in both oil and fibre and consequentially have the lowest starch content of any major cereal derivative.

The choice of the main structure-forming material will depend on several factors based on price and performance. The performance will vary according to composition and compatibility of the non-starch components with the recipe. These factors will be considered later in the sections covering the fillers, lubricants, flavour and colour groupings. However, even the refined starch

Table 2.5 Composition of flours from different cereals as per cent of dry solids*

	Soft wheat	Yellow maize	Rice	Oats
Starch	86	90	90	67
Protein	10	7.5	6	12
Fibre	1.5	0.5	0.5	10
Lipid	1.5	0.6	1.5	8
Ash	0.5	0.5	0.5	2

* Data from samples examined at CCFRA.

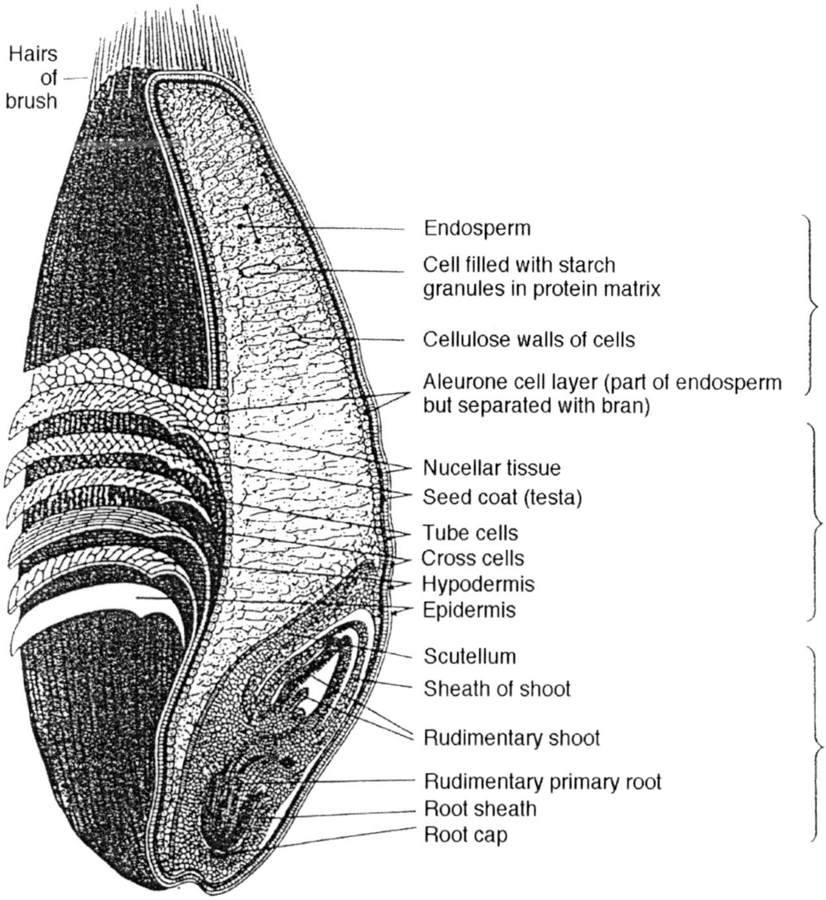

Fig. 2.4 Diagram of the cross-section of a wheat grain showing the major components (Wheat Flour Institute, USA).

component of a cereal may differ from one type to another in a significant manner with respect to performance. Therefore it is necessary to look at the starch component in more detailed form in terms of its composition and physical form.

The physical nature of the cereal flours

Most extrusion manufacturers use flour for their manufacturing processes because it is usually the cheapest form of raw material. Milling cereal grains with either hammer or roller mills to fine particle sizes forms flours that are suitable for extrusion cooking, which is not itself designed as a mill. The small particles of flour may vary in their physical performance in the extruder according to their nature.

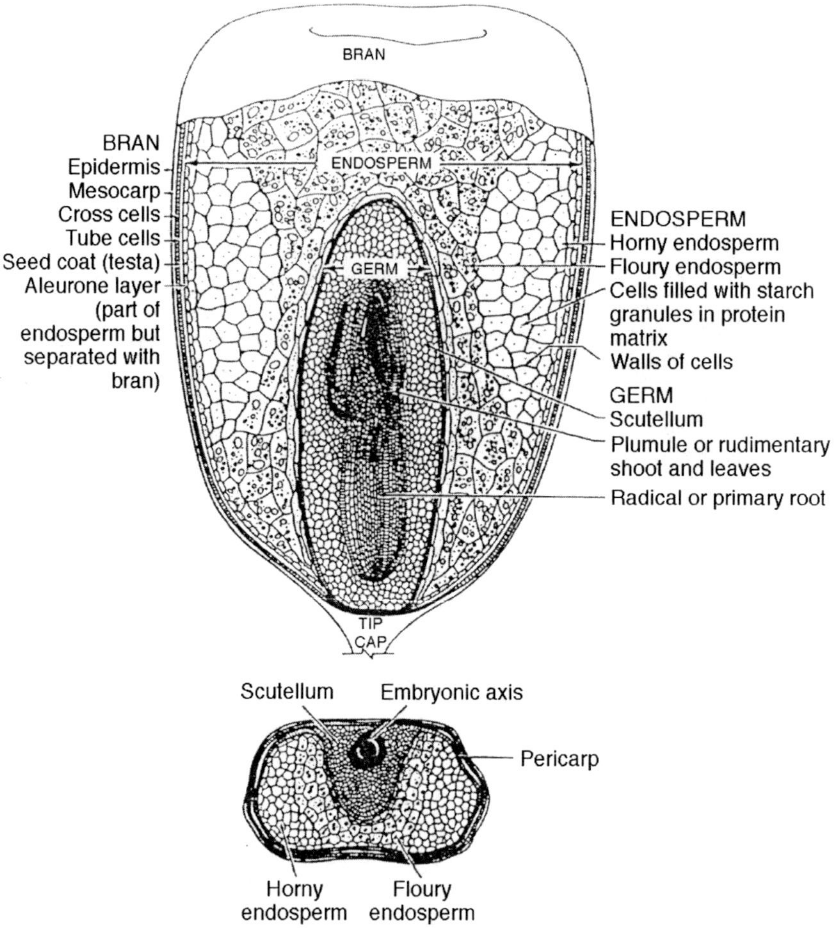

Fig. 2.5 Diagram of the cross-section of a maize grain showing the major components (Corn Refiners Association, USA).

(a) Soft or floury endosperm
These are varieties of wheat, rye, barley and portions of the maize grain from the inner endosperm which are soft and floury. In this material the starch granules and protein layer are only loosely bound together and the endosperm is broken down easily on milling to provide a mixture of separated starch and protein bodies. In the extruder soft flour will create less mechanical energy between its particles and require less mechanical energy to process through the same screw configuration. However, it will create less heat and may have a longer time before melt formation and less time for the transformation of the melt in the shearing section.

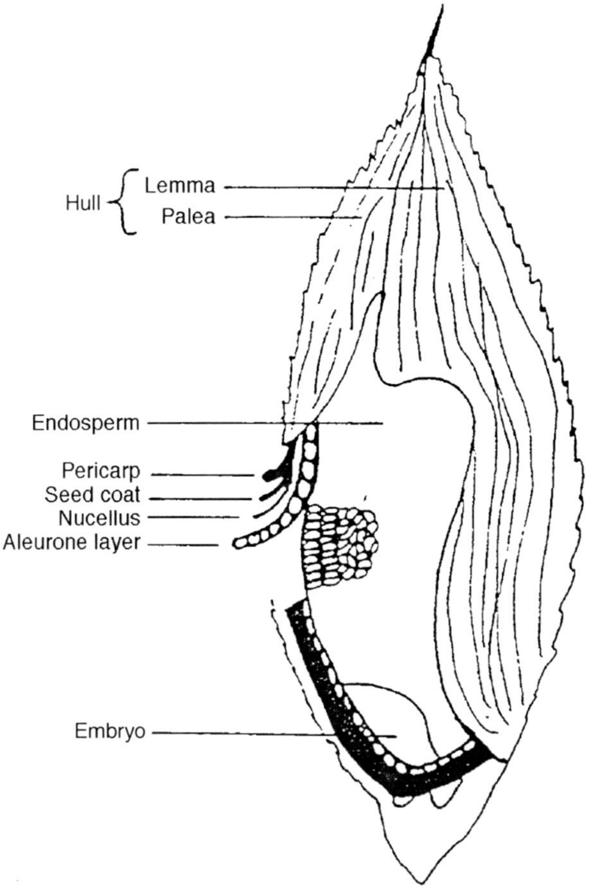

Fig. 2.6 Diagram of the cross-section of a rice grain showing the major component.

(b) Hard and vitreous endosperm
In certain cereals such as rice, hard wheat, durum wheat, vitreous flint maize and some varieties of barley there is a strong bonding between the starch granules and the protein layers. This forms a hard particle of flour that requires more energy to breakdown and will create more heat in the extruder. If the flour particles are small in the extruder they will melt quickly and give extra time for development in the shearing zone of a screw. However, if they are large they may take a longer time to melt because of the heat transfer requirement and the fact that they are strong and do not break down easily as in the case of the soft flours.

Flours may be formed from blends of grains that may contain both hard and soft types in those cereals that have both types, such as wheat and maize. This allows a constant mixture to be used to run a process and to set up the raw material to suit the product. For example, if high expansion is required in a low moisture product, finely milled forms of the harder endosperm types will give excellent results. If the product requires a low to medium expansion some of the hard material may be replaced by soft flour, and for a dense low expansion in a product such as breading crumb, soft flour may be used.

Sources of starch: pregelatinised flours from potato and maize
The raw materials derived from potatoes have traditionally been precooked either on roller dryers or in a granulation drying plant. The roller-dried flake contains pregelatinised starch granules and cell debris. Its starch granules are free to hydrate in the extruder and vary in their degree of damage. In the screw system of an extruder it behaves as a soft viscous mass and is broken down easily to disperse the starch.

Potato granules are made by cooking potatoes as slices or cubes, so that they retain some of their cellular structure. Consequently, some of the pregelatinised starch is still trapped with the potato cells and gives a firmer texture. The moist potato is cooked to gelatinise all the starch and dried carefully by an add-back process to lower their moisture content by adding dry material into the wet potato. Sulphur dioxide and monoglyceride may be added during the cooking stage to control the quality. After equilibrating the mass to moisture < 35–40%, it is cooled to 20–25°C and stirred to form individual potato cells with some retrogradation of the starch. Drying is completed by an airlift to 12% moisture and a normal hot air system to 5–7% moisture without changing the granular structure

The ancient culture of the Aztecs in Mexico produced pregelatinised forms of maize by cooking whole grains with 0.1–0.2% limewater. They produced precooked grains and called the process nixamalisation. In current practice the processing may be varied to give moisture levels in the cooked grain from 35 up to 50%. After cooking and steeping, the swollen grains are washed and milled to dough on a stone mill and sheeted out to form tortilla dough. The dough can be dried and ground to form masa flour. This flour is coarse and may have different particle sizes according to the milling process and the different sizes may be used in different types of product. Attempts have been made to produce similar products by extruding the maize with some lime from a single or twin screw machine but the hot water cooking system is used for the majority of the supplies to the snack industry.

2.3 Examples from Group 1: structure-forming materials based on protein

Proteins are formed from chains of amino acids and have a wide range of physical sizes and forms in native raw materials. The Osborne classification describes the following groups found in cereals:

Table 2.6 Examples of protein-based products

Product group	Product type	Protein source
Human	beefburger	soya/wheat gluten
	dehydrated meals	soya/wheat gluten
	pot noodle	soya
Petfoods	canned chunks	soya, wheat gluten, fava bean
	dry chews	soya, wheat gluten

- albumins: small globular water-soluble proteins
- globulins: small globular proteins soluble in dilute saline
- gliadins: medium-sized proteins soluble in 40% alcohol
- glutenins: large polymeric proteins insoluble in 40% alcohol.

Some of these groups are found in other major protein sources such as oilseed proteins.

In extrusion the protein types that have been found to form a continuous structure are globular proteins from the oilseeds such as soya rape, sunflower and cottonseed and the separated gliadin/glutenin mixtures washed out of wheat flour. The mechanism for structure creation with the proteins is similar to starch in that the proteins must be dispersed from their native bodies into a free flowing continuous mass. This requires a special set of conditions in which water levels are from 35–40%, protein concentrations > 40%, temperatures > 150°C and high shear forces. In the fluid state at high temperature the protein is amorphous and disrupted by the mixing action of the extruder. In order to form a textured structure it must be allowed to flow through a die at a temperature of 120–130°C so that its viscosity increases and the fluid forms laminar flowlines. The texturisation occurs between the molecules as they flow together in the streamlines by some form of laminar cross-bonding. The evaporation of water in the dough mass creates gas bubbles that form alveolar structures held in place by the cross-bonding in the layer.

Examples of protein-based products are shown in Table 2.6.

2.4 Examples from Group 2: dispersed-phase filling materials

The extensive studies of the melt extrusion of synthetic polymers such as polystyrene and polyethylene have shown that their rheological characteristics can be modified by the addition of dispersed particles of colour and carbon black. These materials have been used to change the characteristics of plastics. They occur as dispersed particles within the continuous phase of the plastic and are generally designated by the term filler.

In the food and feed extrusion recipes the presence of fillers is widespread and has been shown to have a similar effect on rheology to the materials added to synthetic plastics. If sections of an extrudate are viewed under a light microscope with appropriate staining techniques, the dispersed materials can be observed as small particles within the continuous phase. In food and feed systems there are four main types of dispersed materials: proteins, starches, fibrous polysaccharides and oils. Some other very small particles may also exist, such as minerals, which will be dealt with in section 2.7.

2.4.1 Proteins

In any starch-rich recipe the different types of protein present will form dispersed phases. The water-soluble proteins such as albumins (egg and whey proteins) will be denatured by heat and coagulated into a soft hydrated mass. This will be macerated to very small pieces, in the range 1–20 μm by the shearing action of the extruder.

The globular proteins from the oilseeds will occur as relatively large particles in the raw materials, but they will also be macerated down to the same general size range by the extruder. These proteins hydrate and form viscoelastic dough pieces by agglomeration with the water. In the shearing zone the protein pieces are macerated in the shearing fields between the screws and the other materials.

The cereal proteins, which are predominantly of the prolamin and glutenin types, also form viscoelastic dough which is macerated to small particles $< 20\,\mu$m in maximum dimension.

Meat proteins from macerated muscle proteins form more resistant particles, which may resist the shearing forces of the extruder and retain the size ranges that existed in the preparation added to the machine.

2.4.2 Starches

In many extruded products the starches will be melted and dispersed into the continuous phase, but in low shear systems some of the granules will remain in the form of aggregates and act as dispersed phases. The amylomaize starches have a more strongly-bonded structure and melt and soften at a much higher temperature than normal starches. If they are added to low shear systems or to gelatinised starch extrusions they may remain intact in the extrudates as dispersed phase material.

In texturised protein products all types of starch will remain in a dispersed state unless the concentration of protein falls below about 35–40%. They can be detected as granular bodies in the protein continuum by staining with iodine.

2.5 Examples from Group 3: ingredients that act as plasticisers and lubricants

2.5.1 Water

In the mass of raw materials the polymer solids create a high viscosity phase, which becomes a glass at ambient temperatures unless there is some liquid material to solvate the polymers and allow them to move freely in the mass of solids. In food and feed extrusion cooking it is water that hydrates and solvates the starch and protein polymers. At levels of > 10% there is sufficient water for the polymers to begin to move and slide across each other and the physical nature of the extrudates changes from a glassy state to a viscous elastic fluid. At this water level the energy expended by the screw conveying and shearing the fluid is very large and the mass heats up rapidly. The extruder also degrades the polymer by mechanical shear and heat. As the water level is increased the viscosity falls and the fluidity of the mass increases and less mechanical energy is expended. At about 25% moisture content in a starch system the heat input is supplied almost entirely by viscous dissipation and friction, but at higher moisture levels barrel heating is required to reach operating temperatures of > 120°C.

The presence of an excess of water in starch, so that there is free water, allows the starch granules to swell after their crystalline structures have been melted. This weakens the granular structures and greatly facilitates the dispersal of the starch polymers. For normal starches the amount of water required to reach this level is about 30% of dough mass.

2.5.2 Oils and fats

Oils and fats have a powerful influence on extrusion cooking processes by acting as lubricants between the particulate matter and the screws of the extruder. Both materials can be described as lipids but fats are those materials, which contain crystalline material at ambient temperatures and appear to be solids. In the extruder they all become liquid at temperatures > 40°C and so function as liquid oils during critical stages of the process. They become mixed with the other materials and are rapidly dispersed to a fine oil droplet size < 1–5 μm and are trapped in the continuous phase. Most raw materials contain some forms of lipids that serve to act as lubricants but there are a few materials, which have only traces of lipid. These serve to demonstrate the function of the oils and fats. If potato or pea starch is extruded at low moisture ($\subseteq$ 16%) the polymers are subject to a high frictional heating effect on the metal surfaces of the screws and barrel, and are degraded to sticky brown gums. Addition of 0.5% of an oil, such as soya, palm, rape or maize, to the extrusion prevents this degradation and allows the starch to be extruded in same way as wheat or maize starches, which contain similar levels of natural lipid. In the extrusion of cereal starches the mechanical energy input is reduced as oil is added to the recipe. Oil reduces the friction between particles in the mix and between the screw surfaces and the fluid so that the forces acting on the starch granules are reduced and they may take longer to disperse in the shear zone. At oil levels > 2% the starch granules may be melted during the extrusion but not dispersed. Consequently the extrudate will be cooked but have no expansion. This phenomenon is observed when a flour from oats is extruded because oat flours may contain > 8% oil content. The effect of oil on starch dispersal is more noticeable at low moisture

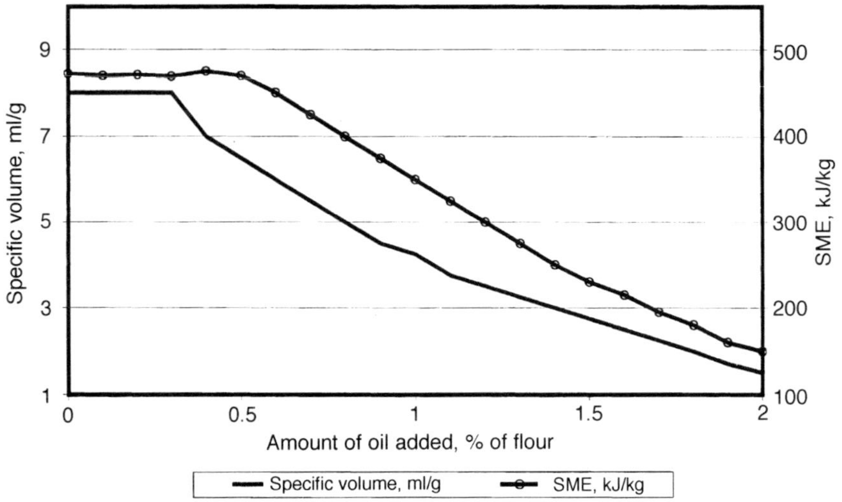

Fig. 2.7 Graph of the effect of oil addition to the extrusion cooking of wheat flour at 16% moisture at high shear/temperature conditions.

up to 30% because in this range the dispersal of starch is achieved by mechanical shear. Above 30% moisture the effect of the free water in swelling the starch reduces the negative effect of the oil and allows increased expansion of the extrudates.

In Fig. 2.7 the effect of addition of oil to wheat flour during extrusion at 16% moisture and a high specific mechanical energy input of 470 kJ/kg is shown. The initial high expansion of 8 ml/g in the extrudate is lost as the mechanical energy input is reduced by the addition of oil. In this simple system it takes only 2% added oil to reduce the expansion to < 1.5 ml/g.

2.6 Examples from group 4: soluble solids

The most common forms of soluble solids used in extrusion are small carbohydrates and salt. They dissolve in the aqueous phase and form a more viscous plasticising fluid but have little effect on most of the structure-forming biopolymers at low levels of addition (< 5%). However, they can reduce the level of starch in a recipe by their very presence in the melt fluid. For example if sugar is added at 10% of the dry recipe for an extruded product, using 15% total water for the extrusion, the maize is reduced to 60% in a simple replacement (see Table 2.7). This means that starch in the maize (shown in brackets) is plasticised by 15 parts of water in each case and the ratio of water to starch increases from 1: 3.5 to 1: 3.0 and in terms of the volume of fluid from 1: 3.5 to 1: 2.25.

The addition of the sugar has increased the volume and diluted the starch in the fluid so that its mechanical reaction to the compression and shearing of the

Table 2.7 Recipes showing effects of adding sugar

Recipe	A	B
Maize	70 (52.5)	60 (45)
Sugar	0	10
Salt	2	2
Flavour	0.5	0.5
Water	5.2	6.6

screw is reduced and the specific mechanical energy input falls. This reduces temperature and the amount of starch degradation so that the extrudate is less well expanded. A solution to this problem is to reduce the water added to the process, which restores the relative concentration of starch to water.

The sugar will change the textural character of the extrudate when it is dried and may form a less porous glass in the cell walls to give a crunchier bite.

2.7 Examples from Group 5: nucleating substances

The principle of nucleation in extrusion was recognised fairly early in the development of the technology, but it was later in the 1980s that the effect of common substances such as bran to act as nucleants was first reported.[1] It was shown that bran particles survived the extrusion process and increased the number of gas cells found in extrudates. It was also shown that the nucleation of gas cells in the expanding extrudate changed the nature of the expansion from anisotropic with die swell to a more isotropic nature.

Later studies at CCFRA and in other reports showed that materials such as calcium carbonate and insoluble calcium phosphate salts nucleated gas cells very well at lower concentrations than the bran. In parallel work in expanded wheat flour for packaging it was shown that talc was equally good for increasing the fineness in the texture of extrudates. It was shown that the nucleation phenomenon requires insoluble particles.

2.8 Examples from Group 6: colouring substances

The colouring of extrudates may occur through the addition of stable colour compounds, or by the generation of colour from precursors in the ingredient mix. There may be some natural colours in maize, which can be yellow, red or blue, but most base materials have only pale coloration and are usually off-white or slightly yellowish.

The addition of colours to the dry mix for the extrusion may be a successful route to colour the extrudate provided the colour is heat stable under the severe extrusion conditions. A number of synthetic colour additives are available for

Table 2.8 Stability of natural colours

Colorant	Destruction in extruder of pure compound, %	Destruction of coated material, %	Half-life in days storage best conditions
β-carotene	75	10–45	< 100
Canthaxthin	30–35	20	180–200
Annatto	10–20	10–20	> 400

use in foodstuffs with rules covering their usage in individual communities. There are a number of colours that have sufficient heat stability to be used in products where the temperatures are not excessive, i.e. > 150–160°C.[6] These colours may be used at 150–300 ppm in the food.

Natural colours in maize and those added from other sources tend to be heat labile and disappear at high temperatures. For example, yellow maize used in a 45 s snack extrusion process will produce yellow snacks at temperatures of 120°C, but lose most of this colour at 150°C. The addition of materials such as β-carotene, Canthaxthin and Annatto have been studied in detail for the effects of extrusion cooking and subsequent storage on the decay of the colour (see Table 2.8). Annatto is the most stable material and is a practical material for use in extruded foods.

The formation of colour in extruded foods may occur by the Maillard browning reaction between reducing sugars and amine groups on amino acids and peptides. The subsequent condensation reactions and formation of polymeric phenolic compounds leads to the development of pale reddish brown colours, which darken with increased heating time. In a wheat product they may form from the flour itself as a red-brown colour, whereas with maize they blend with the natural yellow to form a range of yellow to orange colours. Normally for the development of satisfactory colours, materials such as milk powder or whey are used to provide the reducing sugar and amine groups.

2.9 Examples from Group 7: flavouring substances

The flavouring of extruded products follows a similar pattern to colouring. A product may develop flavour by thermal reactions between flavour precursors in the mix or be flavoured by adding synthetic or natural flavourings. The addition of flavouring is usually carried out on the dry extrudate by spraying or dusting, because of the changes caused by the losses of volatile during extrusion.[7]

Studies at CCFRA and the University of Reading have shown that a background flavour can be developed in cereals such as wheat, maize, rice and oats during processing at temperatures of 120–180°C.[8,9] This flavour varies with temperature as the balance of compounds changes. For wheat and maize the most attractive range for flavours is from 130–160°C when the main flavours are similar to gun puffed wheat and popcorn. Rice has a less attractive range of flavours with some volatile sulphur compounds giving chemical aroma like hydrogen sulphide.

Adding reducing sugars and amino acids or peptides to produce a new dominant flavour profile can alter the basic flavour in a feedstock. In trials with precursor levels of 0.6–0.8% of the feedstock, stronger flavours were developed with nutty, biscuity or other characteristics, depending on the combination of reducing sugar and amino acid used.

An important part of the study of flavour generation at CCFRA and the University of Reading was the development of an understanding of the way flavour compounds were retained in the extrudate after being extruded at high temperatures. It was shown that the water soluble flavour compounds are lost by evaporation in decreasing amounts as their boiling point increases. This was similar to earlier findings.[7] Surprisingly the temperature of the extrusion at the die exit had little effect on the loss of any flavour compound between 120 and 180°C.

Aroma compounds were retained in the glassy extrudate after cooling and were stable for long periods. They were released by wetting the extrudate so that the material became fluid and the aroma compound occupied the headspace above the wet material. Analysis of all the compounds present in an extrudate showed that most compounds were released completely by the change.

2.10 References

1. GUY, R.C.E. and HORNE, A.W. (1988). Extrusion cooking and co-extrusion. In: Blanshard, J.M.V. and Mitchell, J.R. (eds), *Food Structure: Its Creation and Evaluation*, Butterworths, London, Chapter 18; 331–49.
2. GUY, R.C.E. and HORNE, A.W. (1989) The effects of endosperm texture on the performance of wheat flours. *Milling, Flour and Feeds*, **182**, Feb. ix–xxii.
3. GUY, R.C.E. (1994) Raw materials. In Frame, N.D. (ed.), *The Technology of Extrusion Cooking*, Blackie, London, Chapter 2; 52–72.
4. BANKS, W and GREENWOOD, C.T. (1975) *Starch and its Components*, The University Press Edinburgh, 307–8.
5. WHISTLER, R.L. and PASCHALL, E.F. (1967) *Starch: Chemistry and Technology, volume II*, Academic Press, London, 654–85.
6. BERSET, C. (1989) Color. In Mercier, C., Linko, P. and Harper, J.M. (eds), *Extrusion Cooking*, American Association of Cereal Chemists, St. Paul, Minn., USA, Chapter 12; 371–86.
7. MAGA, J.A. (1989) Flavour formation and retention during extrusion. In Mercier, C., Linko, P. and Harper, J.M. (eds), *Extrusion Cooking* American Association of Cereal Chemists, St. Paul, Minn., USA.
8. BREDIE, W.L., MOTTRAM, D.M. and GUY, R.C.E. (1998) Aroma volatiles

generated during extrusion cooking of maize flour. *J. Agricultural and Food Chemistry*, **46**, 1479–87.

9. BREDIE, W.L., MOTTRAM, D.M., HASSELL, G.M. and GUY, R.C.E. (1998) Sensory characterisation of the aromas generated in extruded maize and wheat flour. *J. Cereal Chemistry*, **28**, 97–106.

3

Selecting the right extruder*

M. N. Riaz, Texas A&M University, College Station

3.1 Introduction and terminology

Basic extruder technology has been used in various forms and industries for many years. New equipment designs have increased the range of extrusion applications in food processing. Today's consumers are demanding a broader selection of foods. Extrusion processing equipment has become the standard in many food industries throughout the world (Riaz *et al.*, 1996).

Food extrusion is a process in which food ingredients are forced to flow, under one or several conditions of mixing, heating and shear, through a die that forms and/or puff-dries the ingredients (Rossen and Miller, 1973). Food extruders can be visualized as devices that can transform a variety of raw ingredients into intermediate and finished products. The cooking temperature can be as high as 180–190°C (355–375°F) during extrusion, but residence time is usually only 20–40 seconds. For this reason, the extrusion cooking process can be called a high temperature short time (HTST) process. It is important to learn extrusion terminology, and to remember that many manufacturers use terms based on their own equipment.

3.2 Function and advantages of extruder technology

Food extruders can perform one or several functions at the same time while processing food or feed (Riaz, 2000):

* This chapter reviews criteria for selecting the right type of extruder for the job at hand. Discussions about specific machinery in this chapter do not constitute endorsement or preference, by the Texas A&M University System or its divisions, of any manufacturer, their products or services.

- mixing
- degassing ingredients
- homogenization
- grinding
- shearing
- starch cooking (gelatinization)
- protein denaturation and texturization
- texture alteration
- enzyme in-activation
- pasteurization and sterilization of food spoilage and pathogenic micro-organisms
- thermal cooking
- shaping products
- expansion, puffing
- agglomerating ingredients
- dehydration
- unitizing

Extrusion technology provides several advantages over traditional methods of food and feed processing, including the following (Smith, 1971 and Riaz, 2000, with modifications):

- options for processing a variety of food products by changing a minor ingredient and/or processing conditions on the machine
- different shapes, textures, colors, and appearances obtained by minor changes in hardware and processing conditions
- energy efficient processing, and often lower in cost compared to other options
- availability of automation with most new extruders, which can increase productivity
- improved product quality over other processes because cooking is done in a very short time and less destruction of heat sensitive ingredients occurs
- easy scale-up of extrusion processes from pilot plant to commercial production.

3.3 Selecting an extruder

Many options, which sometimes confuse buyers, are available in the marketplace when selecting extrusion systems for product development. For example:

1) Is a single- or twin-screw extruder required?
2) Should it be a 'wet' or 'dry' extruder?
3) Should it have internal steam locks or a single face die plate?
4) Should it have continuous or interrupted flights?

Appropriate selection depends on several factors:

- Physical and sensory properties of the end product.
- Formula ingredients: their physical nature (i.e., will the product utilize high levels of fresh meat?), moisture content; constant availability or seasonal ingredients; and substitute ingredients that may be used occasionally.
- Kind of product to be extruded? Food grade or feed/pet food? Should each piece be multi-colored or center-filled? Is the shape general, exotic, or detailed? What is the target bulk density? In case of feed, how much fat needs to be added in the formula? How much can be applied to the surface?
- What is the production rate? The size of an extruder depends on market size since extruders function best operating at full throughput per hour.
- What is the source of energy? Is steam or electricity (for product heating) more economical where the extrusion plant will be built? If it is a small operation in a developing country, would a tractor power take-off drive be more suitable?
- What about capital availability, and the recovery date target? Would a used extruder fill the need better for a start-up operation?

Choosing the proper extruder configuration is critical for successful extrusion. The extruder manufacturer should be able and willing to assist in tailoring screw, barrel, and supporting equipment configurations for processing specific products. All these factors should be considered when deciding which kind of extruder best fulfills needs. The four most commonly-used types of cooking extruders currently are: single-screw 'wet' extruders, single-screw 'dry' extruders, single-screw interrupted-flight extruders, and twin-screw extruders. Once the appropriate extruder is selected, it must be assembled correctly and then adequately maintained. Operator training is important, and the supplier of extrusion equipment must be able to provide this service.

3.4 General design features

All extruders consist of a screw(s) which conveys the premixed ingredients through the barrel. Regardless of whether the machine is single- or twin-screw type, several principles apply to all. Screws generally are suspended only from the drive end of the barrel, and rest on the product at the exit end. As a result, the greatest stress and wear on the screw and barrel occur at the exit, and these parts need refurbishing or replacement first. But, complete screws and barrels of even small commercial size extruders are heavy and difficult to transport and set up in lathes or surface machining equipment. Except for very small or old extruders, both the screw and barrel are segmented. The screw typically consists of a shaft that is splined, equipped with a keyway, or hexagonal shape onto which various elements (flight sections, flight 'worms' of different design, and shearlocks/steamlocks) slip before being tightened in place. In twin-screw extruders, each screw consists of modular components too. This design has two major advantages:

1) the elements can be arranged in a variety of configurations as needed for specific applications
2) the worn exit segment can be replaced as needed, or moved back on the shaft to a position where its increased clearance with the barrel is less critical.

In addition to segmented barrel sections, which often have liners that can be replaced as wear proceeds, provisions must be made to keep the product from turning with the screw. Screws act as positive displacement pumps in twin-screw extruders, and the barrel wall typically is smooth. In the intermeshing co-rotating design, each screw wipes the other in moving product forward; in the intermeshing counter-rotating design, the screws jointly squeeze the product forward. Other provisions must be made for moving product forward in single-screw extruders. The oldest design solution was introduced in meat grinders in the latter 1800s. Rifling or parallel groves were cut (more often cast) into the barrel. Both 'wet' and 'dry' single-screw extruders include this feature. Because the barrels and screw flight sections are segmented, a ring-like 'steamlock' or 'shearlock' can be placed between each section, turning the previous section essentially into a pressurized mixing-shearing-reactor cell. Typically, clearances between the 'lock' and the barrel wall decrease as the product is conveyed forward, resulting in zones of increasing pressure. The second design solution to prevent the product from spinning with the screw was borrowed from the Anderson continuous oil screw press, invented at the end of the 1800s. Instead of the screw segments aligning to form a continuous forward conveying flight, space was intentionally left between the flights, giving rise to the term 'interrupted flight'. The barrel inside this type of machine is smooth walled, but 'shearing bolts' protrude through the barrel wall into the space between the flights. As needed, a hollow bolt can replace a solid bolt and convey steam into the product during processing. The die plate at the discharge end of the extruder is the only restriction to product flow and, conceptually, the entire barrel is one reactor cell.

3.5 Segmented screw/barrel single-screw 'wet' extruders

A typical drawing of the single segmented wet extruder is shown in Fig. 3.1. Segmented screw/barrel single-screw extruders are the most widely applied cooking extrusion design in the food, pet foods and feeds processing industries. 'Wet' means that steam and water can be injected into the barrel during processing. Typically, the barrels of these machines are also equipped with heating and cooling jackets. They process more tonnage of extruded products than any other extruder design. The products produced range from fully cooked, light density corn snacks, to dense, partially cooked and formed pastas (Rokey, 2000). They are the focus of discussion in this section.

A typical single-screw extruder consists of a live bin, feeding screw, preconditioning cylinder, extruder barrel, die and knife. The live bin provides a

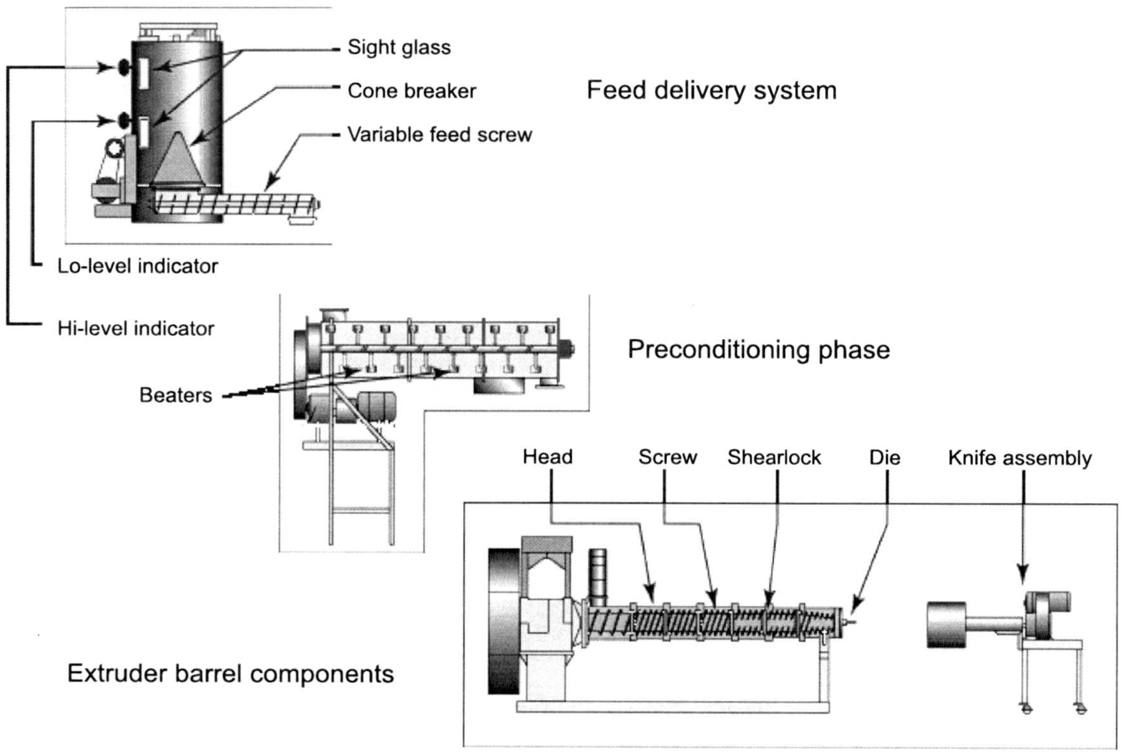

Fig. 3.1 Single-segmented wet extruder.

buffer of raw material so the extruder can operate without interruption. Typically, the height of raw material in the bin is maintained within defined limits by high and low sensors which activate a conveyor supplying the bin. The bin is designed to prevent bridging of its contents and blocking the feed screw leading to the preconditioner. Speed of the feed screw to the conditioner or extruder must be variable to ensure a continuous uniform supply of raw material, which, in turn, leads to consistent and uniform operation of the extruder.

Because single-screw extruders have relatively poor mixing ability, they are usually supplied with premixed material which often has been preconditioned with added steam and water. Generally, preconditioning prior to extrusion enhances extrusion processes which benefit from higher moisture content and longer equilibration time. Preconditioning of the raw material typically improves the life of wearing components in the extruder by several fold. Although the weight of ingredients in the extrusion system is increased, preconditioners are relatively inexpensive to build for the volume they hold and time added to the process for preconditioning. Product quality can be improved greatly by preconditioning the raw ingredients.

The single-screw extruder barrel assembly is composed of a jacketed head, a rotating extruder shaft which carries screws and shearlocks, a stationary barrel housing, a die, and the product cut-off knife. The screws are the key element of the single-screw extruder and their geometry influences performance of the extruder. The barrel bore may be uniform in diameter from inlet to discharge; it can be tapered, decreasing in a bore diameter from inlet to discharge; or it can be of uniform diameter with the final segment of the barrel being tapered or decreasing in diameter. A screw configuration consisting of a variable pitch, constant depth, increasing root diameter, increasing number of flights, shear-locks, and decreasing end diameter is most frequently used in the food industry.

A single-screw barrel can be divided into three processing zones: feeding zone, kneading zone and the final cooking zone (Mercier *et al.*, 1989). The feeding zone generally has deep channels which receive the feed. The preconditioned or dry material entering this zone is conveyed to the kneading zone. Water may be injected at this point to help develop a dough and improve heat transfer in the extruder barrel. As the material is conveyed into the kneading zone, its density increases because of water and steam addition. Screw pitch in this zone decreases and the flight angle also decreases to facilitate mixing and a higher degree of barrel fill. This zone applies compression, mild shear and thermal energy to the feedstock, and the extrudate begins to lose some of its granular definition. By the end of this zone, the feed material is a viscoamorphic mass at or above 100°C (212°F) (Faubion *et al.*, 1982). The reduced slip at the barrel wall prevents the food material from turning with the screw, referred to as 'drag flow' (Miller, 1990). A continuous screw channel serves as a path for 'pressure-induced flow' because the pressure behind the die is much higher than at the extruder inlet. 'Leakage flow' also occurs in the clearance between the screw tip and the barrel wall. The flight of the screw may be interrupted in this area to further increase mixing via leakage flow (Rokey, 2000). The mechanism of shear begins to play a dominant role because of the barrel fill in this zone. Steam and water can be injected in the early part of this zone. Steam injection increases thermal energy and the moisture content of the extrudate. As the extrudate moves through the kneading zone, it begins to form an increasingly cohesive flowing dough mass, which typically reaches its maximum compaction. The material exhibits a rubbery texture similar to a very warm dough. At this stage, the material enters the extruder final cooking zone. Screw flights in this zone are typically shallow, and have a short pitch. The function of this zone is to compress and pump the material in the form of a plasticized mass to the die. Temperature and pressure typically increase very rapidly in this region because of the extruder screw configuration. Shear is highest in this zone, and product temperature reaches its maximum and is held for less than five seconds before the product is forced through the die (Harper, 1978). The product expands as a result of moisture vaporization as it exits through the die into a region of lower pressure. The extruded material can then be cut into desired lengths by the knife attachment.

3.5.1 Application

The first major commercial application of the single-screw extruder in the food processing industry was conversion of semolina flour into pasta using solid screws. This low-shear, low-temperature-forming process first found commercial production in the 1920s and 1930s, and remains a standard process although equipment has improved (Huber, 2000). Several new developments in the single-screw extruder have further increased its efficiency and versatility. A brief list of the products made by single-screw extruders includes:

- direct expanded corn snacks
- texturized vegetable protein
- ready-to-eat breakfast cereal
- production of full fat soy
- pet foods
- floating and sinking aquatic feed
- production of baby foods
- rice bran stabilization
- precooked or thermally modified starches, flours and grain
- breading.

3.5.2 Pros and cons

Single-screw segmented 'wet' extruders are easy to operate and less training is required for the operators. Single-screw machines cost about half the price of twin-screw extruders and maintenance costs are lower. Fewer complications exist in assembling screw configurations as compared to twin-screw extruders, because intermeshing between two screws is not required. 'Wet extruders' have higher capital investment than 'dry extruders', but usually have lower operating costs. Wet extruders have higher capacities than dry extruders due to large drive motor requirements per unit throughput on dry extruders. 'Wet extruders' yield superior shaped products compared to 'dry extruders', because of more processing control.

Since the single-screw extruder has only one shaft, it will not self-clean as completely at the end of the operation. However, if formulas of pet food or feed products are nearly similar, a prepared second swing-away die plate assembly can be mounted at the discharge end. Usually, the extruder can be stopped, the die plate loosened, swung out, and replaced by another with a very short downtime. If not allowed to cool, the extruder can be started up. Opportunity may exist to add slowly up to 10% of the start-up product as rework in the succeeding run. Also, increased water or oil and ground corn or cracked soybeans may be added to help clear the cooling extruder at shutdown. Typically, production extruders operate 24 hours a day, and shutdowns for cleanup do not occur except for required sanitation when meat ingredients are used in food products. Recipes containing more than 12% internal fat may cause slippage inside the barrel, resulting in less shear, pressure and cooking of the product. Ingredients' grind range (sieve size) may be limiting. Very fine powder will not feed in this type of extruder as well unless preconditioned, and very coarse material will not cook properly. Final particle diameter sizes can also be a factor, fish feeds smaller than 1.5 mm may be hard to produce.

3.6 Dry extruders

The term 'dry' extrusion means that this type of extruder does not require an external source of heat or steam for injection or jacket heating, and all product heating is accomplished by mechanical friction (Said, 2000). This type of extruder was developed initially for processing whole soybeans on the farm. A typical dry extruder is shown in Fig. 3.2. Dry extruders can process ingredients which have a wide range of moisture contents, i.e. 10–40%, depending on the premixed formula. If the ingredients have sufficiently low initial moisture content, drying of the product after extrusion cooking may not be necessary. Moisture loss in dry extrusion is in the form of steam flash-off at the die, and the extent depends on initial moisture in the ingredients and product exit temperature. Dry extruders have the option of water injection during extrusion. Usually starchy material requires some moisture in order to gelatinize.

Considerable advances have been made in the design of dry extruders and their components. Recent research has shown that efficiency/throughput of the extruder is almost doubled if the starting material can be preconditioned with steam and water. Longer barrels will work much better in some applications than the shorter barrel formerly used for soybeans.

Dry extruders are single-screw extruders with screw segments and steam-locks (choke plates) on the shaft for increasing shear and creating heat. When material moves through the barrel, and comes up against these restrictions, it is unable to pass through, pressure increases, and a back flow is created. Usually

Fig. 3.2 Typical dry extruder.

these restrictions are arranged in such a way that they increase in diameter toward the die end of the screw to create more pressure and shear as the product reaches the die. This build-up of pressure and temperature, together with the shear stresses developed, plasticizes the raw materials into viscous paste or puffed shapes, depending on the raw material. There is no basic difference between the above and the 'wet' extruders, except that more shear occurs in dry extruders to create heat.

In dry extrusion, pressure and temperature are at their maximum just before leaving the die. The die design and opening also play very important roles in pressure build-up. The cooking range in a dry extruder can be 82–160ºC (180–320ºF) with very high pressure. As soon as the material exits the extruder dies, pressure is instantaneously released from the products, causing the internal moisture to vaporize into steam and making the product expand and results in sterilization of the product.

3.6.1 Applications

Dry extruders can be used for food, feed and recycling of food and feed by-products. A major use of the dry extruder is in preparing oilseeds for screw pressing of oil – primarily soybeans and cottonseed, although they have been applied to sunflower, peanut and canola seed processing. In the process, soybeans and cottonseed are extruded using a dry extruder, followed by pressing in a parallel bar screw press to remove the oil. Extrusion prior to screw pressing greatly increases throughput of the expeller over the rated capacity. Oil and meal produced by this method are remarkably stable because extrusion also releases natural antioxidants in oilseeds. This process is used around the world for processing raw soybeans into full fat soybeans and partially defatted soybean meal. Cereal grain fractions and other starchy raw materials can be preceded by dry extrusion. Applications include processing of:

- cereals and starches
- snack foods and breakfast cereals
- textured vegetable protein
- enzyme inactivation in rice bran
- pet food
- aquaculture feed
- feeds for other animals (pig, cattle, horse, mink)
- recycling wet waste from food, and animal by-products.

3.6.2 Pros and cons

Dry extruders require relatively low capital investment and can be engineered to fit all sizes and types of installations. Less training is required as compared to twin-screw extrusion systems. Dry extrusion is a good choice where steam is not available, although it does have the flexibility of adopting a steam pre-conditioner for applications such as complete feed or shaped products. A typical dry extruder with preconditioner is shown in Fig. 3.3. Dry extruders are able to grind whole soybeans during extrusion, and therefore grinding steps can be eliminated in the case of soybeans.

Dry extruders require relatively high horse power to operate in some applications as compared to other extrusion systems. Higher wear occurs on the screw because of lack of lubrication from steam injection. The final size of the product may be limited to certain sizes, as the high pressure involved with this type of extruder makes it extremely difficult to shape product which is less than 2 mm. They are not as flexible as wet segmented single and twin-screw extruders. Maintenance costs are higher when whole soybeans are ground. Grinding soybeans by hammer mills before dry extrusion may be a more cost-efficient operation. Initial moisture content of the recipe is very critical. Formulas with high fat contents may not cook properly because of product slippage inside the barrel. Highly viscous material is hard to process through this type of extruder compared to twin-screw extruders.

3.7 Interrupted flight extruders

The basic design for most of today's interrupted flight extruders (also called 'expanders') was developed and introduced in the United States by the Anderson International Company (Cleveland, Ohio) as the 'Anderson Grain Expander' in

Fig. 3.3 Typical dry extruder with preconditioner.

the latter 1950s for processing pet foods and other cereal products. Expanders were exported to Brazil for stabilizing rice bran in 1965, Ecuador in 1969, and Mexico in 1970. This design was applied to preparing soybeans and cottonseed for solvent extraction in Brazil in the early 1970s. Brazilian-made expanders were brought back to the United States for processing cottonseed in the late 1970s. An estimated 70% of the domestic tonnage of soybeans and cottonseed processed in the United States is now prepared for solvent extraction by interrupted flight extruders. Currently, machines of similar design are made in the United States, Brazil, India, Switzerland and Germany (Lusas and Watkins, 1990). A typical interrupted flight expander extruder is shown in Fig. 3.4.

An interrupted flight extruder is mechanically different from other extruders, because it was developed from a screw press. Screw presses and interrupted flight extruders are similar in that a revolving interrupted flight pushes the material through a cylindrical barrel and out through an opening at the barrel's end. Rather than round 'shear bolts', the protrusions into the open area between flights are called 'breaker bars' in screw presses. However, a screw press is a more massive and costly machine; it generates more pressure, and it is equipped with a barrel section that allows oil to flow away from the solids (Williams, 2000). Although extruders are often equipped with steam-heated/water-cooled jackets, commercial interrupted flight expanders usually are not jacketed and rely on direct steam injection for supplemental heat beyond that created by mechanical shear of the ingredients.

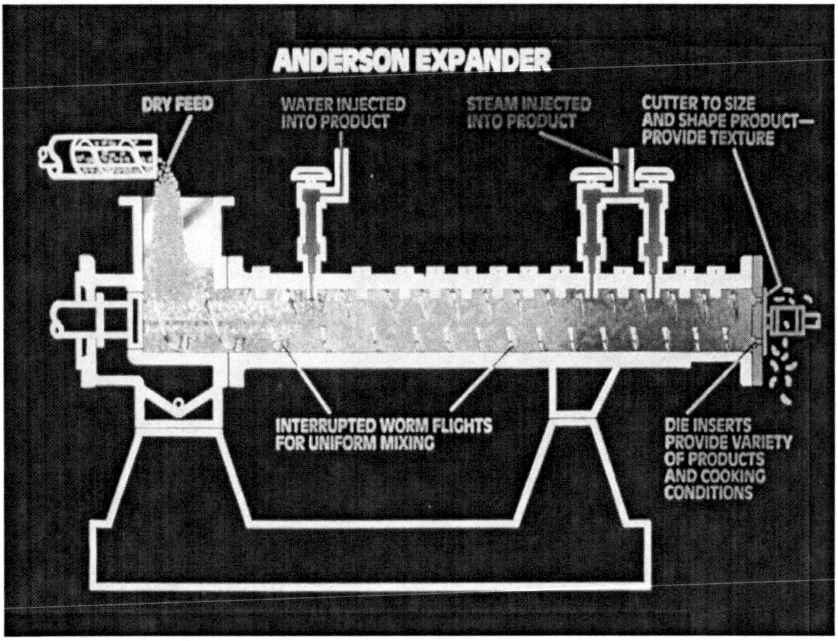

Fig. 3.4 Typical interrupted flight expander extruder.

The expander's internal mechanism consists of a rapidly rotating worm shaft, having individual worms with an interrupted flight positioned inside a smooth walled barrel equipped with removable stationary pins protruding from the barrel and intermeshing with the interruptions of flight. The purpose of the intermingling of the rotating worms with stationary pins is to provide a high shear, turbulent mixing action, which kneads the solid formulation with the injected water and steam to provide rapid and uniform absorption of the injected moisture into all of the solid matter. As the steam is absorbed, it releases its heat of vaporization which elevates the temperature of the ingredients. Frictional heat is also generated by the rapid motion of the flights, further elevating the temperature as they compact and work the mixture subjecting it to increasingly higher pressures as it is forced through the length of the barrel. By the time the mixture reaches the end of the length of the barrel, it is thoroughly cooked and under high enough pressure and temperature (120–150°C; 248–302°F), resulting in much of the moisture flashing off as product exits the expander (Williams, 1988).

3.7.1 Applications

Anderson 6 and 8 inch diameter expanders played a lead role in processing pet foods until modern segmented single-screw extruders were developed. Expanders now play a major role in preparing oilseeds for solvent extraction, and special high shear heads have been developed for shearing oilseeds before screw pressing. Expanders are still used for making pet foods and floating aquatic feeds. Since expanders have less shearing action than dry extruders, soybeans should be ground before entering the machine when making full-fat soybean meal. As a result, less wear generally occurs in expanders than in dry segmented extruders, smaller electric motors can be used, and maintenance costs are lower. A short, interrupted flight shear bolt section is often included in large feed-type expanders, 12 inch diameters and up, used to precondition/precook feed formulas before pelleting. Applications of expanders include:

- full fat soybeans
- oilseeds preparation for solvent extraction
- oilseeds preparation for 'hard pressing' to 5–6% residual oil in the meal
- rice bran stabilization
- pet foods
- aquaculture feed
- feeds for other animals (pig, cattle, horse, mink)
- snack foods
- drying of synthetic rubber and plastic polymers.

3.7.2 Pros and cons

Interrupted flight expanders are relatively less expensive than segmented single- and twin-screw extruders. These are very simple machines, easy to operate, and minimum training is required for the operators. Interrupted flight expanders are very rugged machines and wear parts are easily replaceable. When properly installed, and with consistent ingredients, they produce consistent quality product for long runs. Shaft speed and designs can be varied, which makes this machine applicable to different raw materials. An oil removal cage can be added for high oil content seeds or oil-bearing materials to produce cohesive collets for solvent extraction, and to increase oilseed traction in screw presses. Several screw ends and barrel head designs are available for processing different materials. A wide variety of preconditioners can be adapted to these machines. Less horse power is required than for dry extruders of the same capacity because of lubrication from water (added as steam), but higher moisture products typically require drying after cutting.

Since steamlocks are not present, heating in the barrel from shear is limited to that resulting from the breaker bolts and back pressure of the die plate, and must be supplemented by injected steam. The maximum barrel temperature is limited by the open die area, and by pressure of the available steam and its quality. These machines are less versatile as compared with segmented single- and twin-screw extruders and processing conditions are more difficult to control. They require a finely ground or flaked material feed stock since they do not have the ability to grind themselves. Pellet sizes are limited to larger diameters as compared to twin-screw extruders. High levels of added fat can cause slippage inside the barrel, and it is best to use ingredients containing internal fat when possible.

3.8 Twin-screw extruder

Recent years have seen increasing requirements for new products with intricate shapes and small sizes that are beyond the capabilities of single-screw systems. Twin-screw extruders can fill some of these needs. The term 'twin-screw' applies to extruders with two screws of equal length placed inside the same barrel. Twin-screw extruders are more complicated than single screw extruders, but at the same time provide much more flexibility and better control. Twin-screw extruders are generally categorized according to the direction of screw rotation and to the degree to which the screws intermesh:

1. counter-rotating twin-screw extruders
2. co-rotating twin-screw extruders.

In the counter-rotating position the extruder screw rotates in the opposite direction, whereas in the co-rotating position the screw rotates in the same direction. These two categories can be further subdivided on the basis of position of the screw in relation to one another into: intermeshing and non-intermeshing.

The non-intermeshing twin-screw extruder is like two single-screw extruders sitting side by side with only a small portion of the barrels in common (Clark, 1978). These types of extruders depend on friction for extrusion, just like single-screw extruders. In non-intermeshed extruders, neither pumping nor mixing is positive. Their design does not provide a positive displacement action for pumping the product forward.

A typical twin-screw extruder is shown in Fig. 3.5. In intermeshing twin-screw extruders, the screws partially overlap each other in a figure '8' barrel track, resulting in positive pumping, efficient mixing and self-wiping action (only in co-rotating machines; limited mixing in counter-rotating machines), differentiating these types of extruders from non-intermeshing and single-screw machines. These extruders are like a positive displacement pump, forcing material in the barrel between the screw to move toward the die by rotation of the screw.

Co-rotating self-wiping types of extruders are most commonly used in the food industry. These extruders significantly increased the variety of products that can be made using extrusion technology.

The twin-screw extruder consists of several sub-components very similar to single-screw extruders (live bin, feeding screw, preconditioning cylinder, extruder barrel, jacketed heads and rotating screw). A detailed discussion of these components is presented in the single-screw extruder section. The bearing assembly in the twin-screw extruders is much more complicated because more

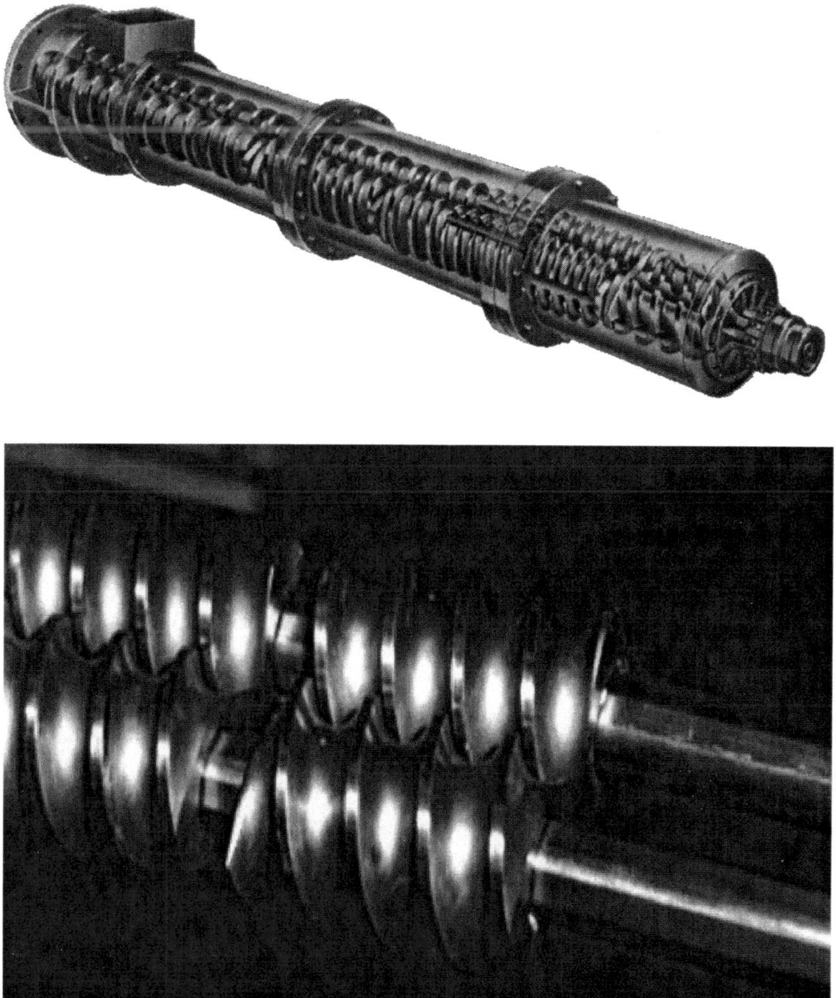

Fig. 3.5 Typical twin-screw extruders.

components (such as drive and torque dividing gears) are required. Twin-screw extruders also have three processing zones, feeding, kneading and a final processing zone very similar to single-screw extruders. These zones are described in the single-screw extruder section.

3.8.1 Applications

Twin-screw extruders got their popularity in the food industry in the mid-1980s–mid-1990s. Originally developed for processing plastics, food companies began using twin-screw extruders for products like sticky caramels and candies, that could not be made with single-screw machines. Very soon, twin-screw extruders become popular with the food manufacturers for specialized food items. Recent improvements in single-screw extruders have made it possible to process several foods formerly made on twin-screw extruders, possibly limiting the market for twin-screw extruders:

1) Variable speed drives (VFD) gave single-screw additional flexibility and are approaching twin-screw
2) Computer control systems
3) Improved gravimetric feed systems and mass flow meters for precise metering of ingredients/recipe components.

Presently, twin-screw extruders are being used for the following different food and feed items.

- co-extruded snacks and other food items
- food gums
- reformed fruit bits and sheets
- topping and bakery analogs
- precooked pasta
- noodles, spaghetti and macaroni
- imitation nuts
- third generation snacks
- bread-like products (crisp bread)
- pastry dough
- pretzel
- ravioli
- texturized vegetable protein (soy)
- wheat gluten textured products
- semi-moist food
- soup and gravy mixes
- sugar crust liqueurs
- pet treats
- three-dimensional snacks
- three-dimensional confections and toffees
- cheese and casein products
- beer powders
- meat analog
- texturized vegetable protein from partially defatted soy flour
- stabilization of rice bran
- multicolor food and snacks
- meat and power bars
- special energy bars with resin filling
- marshmallow products
- cereals and corn flakes

- chocolate filled snacks
- confectionery and other chocolate products
- cocoa and crumb
- cookies and cracker
- corn chips and tortilla
- dairy products
- egg rolls
- jellies
- flavoring
- instant rice
- instant noodles
- beverage powder
- boiled sweets
- candy sticks
- loose fill (packaging material) from starches
- ultra fine aquatic feed
- high fat aquatic feed (salmon)
- premium pet food (with fresh meat).

As we enter a new millennium, these product areas illustrate only a fraction of the almost limitless number of products that can be made using extrusion technology.

3.8.2 Pro and cons

Because of positive pumping action, and reduced pulsating of products exiting the dies, very uniform product lengths and intricate shapes can be produced. Twin-screw extruders can handle very viscous, oily, sticky and wet materials. Usually, they can handle products with greater than 18% internal fat as compared to 17% maximum for single-screw machines. They can produce pellets of less than 1 mm diameter for food or feed applications – this latter capability is important in feeding fish fingerlings. Since the twin-screw extruder contains two shafts, the inside of the barrel is swept clean and the two screws swipe each other. At the end of the operation, some steam and water usually can clean the extruder barrel from the inside. Twin-screw extruders can handle a wide range of the grind sizes of recipe ingredients. Very finely ground ingredients can be fed directly into the twin-screw extruder, as well as very coarse ingredients. Use of a preconditioner is recommended when possible, but is not always practical when processing pregelatinized, or high-amylopectin content ingredients. If the end of the barrel splits the extruder output into two channels, products with pieces of two different colors, or variegated, can be made by injecting color solutions into the channels just before the die. Three-dimensional, or very delicate food products can be processed. Up to 30% fresh meat can be used in recipes processed on twin-screw extruders. Twin-screw extruders are more forgiving to inexperienced operators.

Twin-screw extruders are more expensive (at least double the price) than single-screw extruders. Also, their maintenance costs may be higher. They are relatively more complicated than the single-screw extruder. Operators need to be very careful when assembling a new configuration. It is very easy to mount the segments at the wrong angles, resulting in breakage.

3.9 Single- vs. twin-screw extruder

Single-screw extrusion has been employed successfully in food and feed production over the last 60 years. Because of consumer demands for innovative food products in the market, extruder manufacturers adopted twin-screw extruders approximately 30 years ago. Twin-screw extruders have greater ability and flexibility for controlling both product and process parameters. They are a flexible design permitting easy cleaning and rapid product changeover. Because of the ability to better match the desired shear, the twin-screw extruder has more control over product variability. Screw speed can also be used to compensate for some variations in the properties of the starting material (Hauck, 1988). The twin-screw extruder is a better choice for plants producing a wide variety of high-value products at low volume because the screw speed is such an influential variable.

Single-screw extruders are limited to 12–17% fat level in the formula. Fat above that level reduces friction because of lubrication and does not help the hardware transform mechanical energy into heat for cooking purposes. On the other hand, fat level in the recipe for twin-screw extruders can be as high as 18–22% and still maintain the mechanical energy. This is only possible because of more screw configuration options with twin-screw extruders compared to single-screw machines. In single-screw extruders with the help of steam injection, fat level of the recipe can be achieved as high as 17%, but the same recipes with the addition of steam in twin-screw extruders will process more consistently, which results in better binding of the fat in the product and reduces the leakage of fat from the products during handling and packaging.

Moisture content is very critical during the extrusion process for starch gelatinization and protein denaturation. An average moisture content in a typical formula ranges from 20–28%. Moisture, as steam and/or water, is added to the preconditioner and extruder barrel to help soften raw ingredients and reduce their abrasiveness. Twin-screw extruders have the ability to run under a narrow range or wide ranges of moisture.

Processors should consider twin-screw extruders in the following situations (Riaz, 2000):

- frequent product changeovers
- products with high internal fat content (above 17%)
- addition of a high level of fresh meat in the product (up to 35%)

Table 3.1 Typical process parameters

Process	Temp. (°C)	Max. pressure (bar)	Moisture (%)	Max. Fat (%)	Cook* (%)
Pellet press	60–100		12–18	12	15–30
Expander/ pellet press	90–130	35–40	12–18	12	20–55
Dry extrusion	110–140	40–65	12–18	12**	60–90
Wet extrusion					
Single-screw	80–140	15–30	15–35	22	80–100
Twin-screw	60–160	15–40	10–45	27	80–100

* % cook is starch gelatinization measured by enzyme susceptibility.
** Dry extrusion successfully processes full fat soy (18–20% fat) and other ingredients where final product durability is not a concern.
Source: Wenger Manufacturing Co. Sabetha, Kansas.

Table 3.2 Typical production capacities and costs

Process	Capacity range (ton/h)	Ave. life major wear components (hours)	Average wear cost ($/ton) Full fat complete soy	Average wear cost ($/ton) Diet
Pellet press	2–60	2,300	N/A	1.23
Expander/pellet press	2–40	700*	0.6**	1.14
Dry extrusion	0.5–2	1,000	1.0	1.78
Wet extrusion				
Single-screw	1–22	5,000	0.5	0.89
Twin-screw	1–14	5,000	1.85	2.01

* Reported 9,800 tons before replacement at rate of 14 tons/h.
** Expander only, pellet press not used.
Source: Wenger Manufacturing Co. Sabetha, Kansas.

- uniform size and shapes
- ultra-small product sizes (less than 1.5 mm)
- products made with low density powder
- special formulations.

Table 3.1 describes typical process parameters of different extrusion processes and Table 3.2 describes typical production capacities and costs of different types of extruders.

3.10 Sources of further information and recommended reading

Various information sources on extrusion technology are available, including Internet Websites maintained by the various extruder manufacturers. Magazines which contain articles on extrusion include:

Cereal Food Worlds, American Association of Cereal Chemists, St. Paul, Minnesota
Food Technology, Institute of Food Technologists, Chicago, Illinois
Feed International, Watts Publications, Mount Morris, Illinois
Feed Management, Watts Publications, Mount Morris, Illinois
Petfood Industry, Watts Publications, Mount Morris, Illinois
Feed Tech, Elsevier International, The Netherlands
Food Processing, Putnam Publications, Itasca, Illinois
Prepared Foods, Cahners Food Group Publications, Oak Brook, Illinois
International Aqua Feed, Turret RAI, United Kingdom

Books for further reading include:

CHANG, Y.K. and WANG, S.S. (1999) *Advances in Extrusion Technology. Aquaculture/Animal Feeds and Foods.* Technomic Publishing Co., Lancaster, Pennsylvania.

FAST, R.B. and CALDWELL, E.F. (2000) *Breakfast Cereals and How They are Made*, 2nd edn. American Association of Cereal Chemists, St. Paul. Minnesota.

Focus on Extrusion (Proceedings) (1999) Pet Food Industry. Watt Publishing Co., Mt. Morris, Illinois.

FRAME, N.D. (1994) *The Technology of Extrusion Cooking.* Blackie Academic & Professional, New York.

HARPER, J.M. (1981; 1991) *Extrusion of Foods, Vol. 1 & 2* CRC Press, Inc., Boca Raton, Florida.

HAYAKAWA, I. (1992) *Food Processing by Ultra High Pressure Twin-screw Extrusion.* Technomic Publishing Co., Lancaster, Pennsylvania.

JANSSEN, L.P.B.M. (1978) *Twin-screw Extrusion.* Elsevier Applied Science, New York.

JOWITT, R. (1984) *Extrusion Cooking Technology.* Elsevier Applied Science, New York.

KOKINI, J.L., HO, C. and KARWE, M.V. (1992) *Food Extrusion Science and Technology.* Marcel Dekker, Inc., New York.

MERCIER, C. and CANTARELLI, C. (1986) *Pasta and Extrusion Cooked Foods.* Elsevier Applied Science, New York.

MERCIER, C. LINKO, P. and HARPER, J.M. (1989) *Extrusion Cooking.* American Association of Cereal Chemist, St. Paul., Minnesota.

O'CONNOR, C. (1987) *Extrusion Technology For the Food Industry.* Elsevier Applied Science, New York.

RIAZ, M.N. (2000) *Extruders in Food Applications.* Technomic Publishing Co., Lancaster, Pennsylvania.

WILSON, D. and TRIBELHORN, R.E. (1979) *Low-cost Extrusion Cookers.* Workshop Proceedings. United States Department of Agriculture. Office of International Cooperation and Development., Washington, DC.

WOODROOFE, J.M. (1993) *Dry Extrusion Manual.* Rural Pacific Pty, Ltd., Australia.

ZEUTHEN, P., CHEFTEL, J.C., ERIKSSON, C., JUL, M., LENIGER, H., LINKO, P., VARELA, G. and VOS, G. (1984) *Thermal Processing and Quality of Foods.* Elsevier Applied Science, New York.

3.11 References

CLARK, J.P. (1978) Texturization by extrusion. *J. Texture Studies*, No. 9, p. 109.

FAUBION, J.M., HOSENEY, R.C. and SEIB, P.A. (1982) Functionality of grain components in extrusion. *Cereal Food World*, 27 (5): 212.

HARPER, J.M. (1978) Extrusion processing of food. *Food Technol.*, 32 (7): 67.

HAUCK, B.W. (1988) Is the Single-screw extruder about to become a dinosaur? *Petfood Industry*, 30 (2): 16.

HUBER, G. (2000) Twin-screw Extruders. In *Extruders in Food Applications*, Ed. Riaz, M.N. Technomic Publishing Co. Inc. Lancaster, Pennsylvania. P1-23.81–114.

LUSAS, E.W. and WATKINS, L.R. (1990) Improvement of soybean oil solvent extraction by use of expanders. Paper presented at Extrusion Workshop on Soy/Nutritive Foods, Sponsored by ALADEES (Association Latino-americana de Extrusion de Alimentos Enriquecidos con Soya), Chihuahua, Chihuahua, Mexico. Aug. 23–25.

MERCIER, C., LINKO, P. and HARPER, J.M. (1989) *Extrusion Cooking.* American Association of Cereal Chemists, St. Paul, Minnesota.

MILLER, R. (1990) 'Unit operation and equipment'. In *Breakfast Cereals and How They Are Made*, American Association of Cereal Chemists, St Paul, MN, p. 145.

RIAZ, M.N. (2000) *Extruders in Food Applications.* Technomic Publishing Co. Inc., Lancaster, Pennsylvania.

RIAZ, M.N., LUSAS, E.W. and MOHY-UD-DIN, G. (1996) 'Applications and advantages of extrusion technology in the food industry', *Pak. J. Food Sci.*, 6: 3–4, 85–8.

ROKEY, (2000) Single-screw Extruders In *Extruders in Food Application* Ed. Riaz, M.N. Technomic Publishing Co. Inc. Lancaster, Pennsylvania. P25–50.

ROSSEN, J.L. and MILLER, R.C. (1973) Food extrusion. *Food Technol.*, 27: 46–53.

SAID, W.N. (2000) Dry Extruders. In *Extruders in Food Applications*, Ed. Riaz, M. N. Technomic Publishing Co. Inc. Lancaster, Pennsylvania, pp. 51–62.

SMITH, O.B. (1971) 'Why use extrusion', Symposium on Extrusion: Processes and Product Development. American Association of Cereal Chemists, St Paul, MN.

WILLIAMS, M. A. (1988) Production of Extruded Pet Foods. In *Vegetable Protein Utilization in Human foods and Animal Feedstuffs*, Proceedings of the World Congress . Ed. Applewhite, T. H. AOCS Press Champaign, IL, pp. 171–174.

WILLIAMS, M. A. (2000) Interrupted Flight Expanders-Extruders. In *Extruders in Food Applications*, Ed. Riaz, M. N. Technomic Publishing Co. Inc. Lancaster, Pennsylvania, pp. 63–79.

4

Optimised thermal performance in extrusion

J. Mottaz and L. Bruyas, Clextral, Firminy

4.1 Introduction

The twin-screw extrusion cooking process in the industries producing foods for animal and human consumption has developed considerably over the last 30 years. Twin-screw extrusion technology produces physical and chemical changes in the product through processes of macromixing or micromixing, according to Zoulalian.[1] Mixing involves a substantial transfer of energy that plays a major role in the transformation of the material. The development of new products and the optimisation of existing applications call for an understanding of the transfer phenomena involved. These are essentially transfers of heat, of either thermal origin (where barrel heating and cooling systems are used) or of mechanical origin (the flow of heat generated as the material is sheared by the movement of the screws). The quality of the end product depends upon how the thermal performance of the extruder is controlled and what the thermomechanical history of the product is inside the extruder. Indeed the control of the particular thermal conditions in every process must be perfect if its operation is to be optimised. As a result it is necessary to develop appropriate measurement systems and to be able to predict the thermal performance of the extruder. The parameters it is important to control are the material temperature and pressure, the barrel temperature, and the mechanical and thermal power levels involved. In addition it is useful to have a means of predicting the temperatures of the barrel and the material under given operating conditions. Such a system could be used for controlling the process or for designing an extruder for a particular application.

The purpose of this chapter is to show the importance of heat transfer in the extrusion cooking process, how it can be measured and its importance in extruder design and operation. The first step is to identify the modes of transfer and the different factors concerned. The application of the principles described will be illustrated using the results – both experimental and numerical – of recent research into the understanding and optimisation of the thermal performance of an extruder.

4.2 Heat transfer in extrusion processing

4.2.1 Introduction

It is important to understand and control heat transfer in the extrusion cooking process so that the thermal treatment most suitable for conferring the required qualities on the final product may be used. Heat transfer in extrusion must be tackled at two levels.

- The first approach, considering the extruder as a whole, is global. It involves quantifying the energy levels – mechanical and thermal – involved throughout the extruder: heating, cooling, losses, and changes in internal energy of the material.
- The second approach concerns only the barrel, so is more local. The conversion of the material as it passes through an extruder involves transfer phenomena that vary along its path. The heat transfers therefore have to be analysed in each functional zone of the extruder as described by Colonna *et al.*[2] This breakdown covers solid transport, pressurisation, melting, mixing and/or shearing, molten transport, and the die.

In both approaches – global and local – an energy balance is worked out in order to quantify the power levels and determine certain factors used for understanding the physical phenomena. The energy balances are obtained from heat transfer experiments or models. The energy balance concept involves parameters related to the extruder (its geometry) and to the material (its thermal and rheological properties).

4.2.2 Global thermal balance: extruder

An overall analysis of the extrusion process can be used to correlate the extent of conversion of the material with the power values involved. The aim of this approach is to determine the energy consumption of the process and to quantify the energy transferred to the material. The power values to be considered in an extruder are:

- the mechanical power supplied by the motor: $P_{mechanical}$
- the thermal power supplied by the heating system: $P_{heating}$
- the thermal power absorbed by the cooling circuit: $P_{cooling}$
- the thermal losses to the environment: P_{losses}
- the power absorbed by the material: $P_{material}$

The balance equation containing these power values (unit: W) is written as follows:

$$P_{mechanical} + P_{heating} = P_{cooling} + P_{losses} + P_{material} \tag{4.1}$$

Della Valle *et al.*[3,4] used this method of calculation for estimating the energy efficiency of an extruder. They characterised the extent of conversion of starch in terms of the energy transferred to it. The power absorbed by the material is deducted from the other measured power levels. This represents the quantity of heat necessary for converting the material, which is equivalent to the change in its enthalpy (internal energy of the product depending on the state of the product and the phase change) as it passes through the extruder (sensible heat and enthalpy of fusion). The balance equation shows that the energy transferred to the material is of either mechanical or thermal origin. An extruded product is usually qualified (degree of transformation and cooking of the final product) by the specific mechanical energy, denoted SME. This energy is the ratio of the mechanical power supplied by the flow of extruded material (unit: $kWh.t^{-1}$ or $J.kg^{-1}$). However, certain extruder users prefer to qualify their product in terms of the mechanical torque consumed by the motor (the mechanical power is the product of the torque and the rotational speed of the motor).

4.2.3 Local thermal balance: barrel

Satisfactory thermomechanical conversion of the material requires an understanding of the transfer phenomena taking place within the material and between the material and the barrel. For this purpose, the local modes of transfer must be identified. The thermal changes in the material as it passes along the screws is determined by solving a balance equation based upon a general one-dimensional model of the modes of transfer. This must cover convective heat transfer between the material and the barrel, convective heat transfer between the material and the screw, and a source term (positive or negative energy) within the material. The balance equation is formulated as follows:

$$dH = dq + h_{m/b} \cdot dS_{m/b} \cdot (T_b - T_m) + h_{m/s} \cdot dS_{m/s} \cdot (T_s - T_m) \tag{4.2}$$

where T_b, T_m and T_s are, respectively, the temperature of barrel, material and screw (°C). This equation is adapted to the particular functional zone in question. The first term, dH (W), in the equation is the change in internal energy of the material. This is expressed differently according to the state of the material: powder (or solid state), melting and molten. This term involves thermal properties of the material such as sensible heat, melting point and enthalpy of fusion.

The second term, dq (W), represents the heat sources and/or sinks within the material. In extrusion cooking, we have viscous dissipation (a positive term) and possibly the energy from a chemical reaction (which may be exothermic (positive energy) or endothermic (negative energy)). Viscous dissipation is related to the rheological properties of the material, particularly its viscosity. The extrusion cooking of foodstuffs rarely involves chemical reactions, but there are exception such as casein conversion into caseinate where the reaction is exothermic.

The third and fourth terms respectively represent the heat transferred between the material and the barrel, and between the material and the screws. The concepts of convective heat transfer coefficient, $h_{m/b}$ and $h_{m/s}$ ($W.m^{-2}.K^{-1}$), and transfer area, $dS_{m/b}$ and $dS_{m/s}$ (m^2) are then introduced. Generally speaking, it is assumed that there is no heat transfer between the screws and the material, and in fact very few applications necessitate cooling of the screws.

4.2.4 Fundamental parameters

Analysis of heat transfer inside a twin-screw extruder shows that the physical phenomena involve a number of parameters that must be controlled.

(a) Physical, thermal and rheological properties of foodstuffs

It is important to draw a distinction between the physical and thermal properties and the rheological properties. A detailed review has been drawn up by Della Valle and Vergnes[5] which gives values as well as appropriate methods of measurement.

The physical properties are the moisture content and the specific gravity. The moisture content (MC expressed in %) determines the other properties of the material. One must differentiate between the moisture content in a dry base from the moisture content in a wet base. The former corresponds to the intrinsic quantity of water present in a raw material. The latter is the total quantity of water after water has been added to the material. The specific gravity is used to calculate the volumetric flow and the degree of fill of the screws.

The thermal properties are the specific heat, the melting point, the enthalpy of fusion and the thermal conductivity. The specific heat of the foodstuffs used in extrusion cooking varies from 1500 to 2500 $J.kg^{-1}.K^{-1}$ depending on the nature and state of the material. In the case of a mixture of various ingredients, it is determined using an additive rule that takes into account the percentage and specific heat of each ingredient. The thermal conductivity of foodstuffs lies between 0.1 and 0.5 $W.m^{-1}.K^{-1}$. The temperature and the enthalpy of fusion are necessary for calculating the energy that must be supplied to the material for it to undergo a change of phase. For maize starch, the change in the melting point as a function of the moisture content follows Flory's law (Colonna and Mercier[6]).

The viscosity of the material is the rheological property governing the viscous dissipation generated by the shear stresses. Special instruments such as rheometers are necessary to determine the viscosity. A number of researchers have investigated the rheology of materials used in extrusion cooking: Harper *et al.*,[7] Fletcher *et al.*,[8] Vergnes and Villemaire,[9] Morgan *et al.*,[10] Mohamed *et al.*[11] and Vergnes *et al.*[12]

(b) Convective heat transfer coefficient between material and barrel

The quality of the heat transfer in an extruder depends on the convective heat transfer coefficient between material and barrel. The power transferred is

Table 4.1 Values of the heat transfer coefficient in the transport zone for material in powder state

Author	Heat transfer coefficient	Material
Yacu[13]	30	wheat flour
Tayeb *et al.*[14]	400 to 2000	maize starch
Barrès *et al.*[15]	800 to 1000	maize starch
Chang and Halek[16]	115 to 205	maize flour

Table 4.2 Values of the heat transfer coefficient in the melting zone for single and twin screw extruders

Author	Heat transfer coefficient	Extruder	Material
Yacu[13]	500	twin	wheat flour
Mohamed and Ofoli[17]	191 to 768	twin	soya
Chang and Halek[16]	500	twin	maize flour
Levine and Rockwood[18]	170 to 420	single	hard wheat flour
Mohamed *et al.*[19]	136 to 420	single	soya
Le Roux *et al.*[20]	300	single	pasta

proportional to this coefficient. A comparative study has been done of the various values of this coefficient to be found in the literature. There has been no specific experimental research aimed at quantifying this coefficient. Table 4.1 gives the values of the convective heat transfer coefficient in the transport zone for material in the powder state. For the zones where the material is molten, a larger number of values is given in Table 4.2. The conclusion is that transfer is better first when the material is molten and, secondly, when a twin-screw extruder is used.

We may quote two correlations for the calculation of the heat transfer coefficient in a twin-screw extruder with material in the molten state. Todd[21] has determined a correlation for a polymer. The transfer coefficient $h_{m/b}$ is deduced from the Nusselt number which encompasses the geometrical characteristics of the screws and the physical, thermal and rheological properties of the material. His expression of the Nusselt number (Nu_{Todd}) is as follows:

$$Nu_{Todd} = 0.94\left(\frac{D_{ext}^2 \cdot N \cdot \rho}{\mu}\right)^{0.28} \cdot \left(\frac{C_p \cdot \mu}{\lambda}\right)^{0.33} \cdot \left(\frac{\mu}{\mu_w}\right)^{0.14} = \frac{h_{m/b} \cdot D_{ext}}{\lambda} \quad (4.3)$$

where D_{ext} is external screw diameter (m), N is screw rotation speed (rpm), and λ is measured in $W.m^{-1}.K^{-1}$

The second correlation is that of Skelland[22] which was determined for a heat exchanger with a roughened surface. In view of the geometry of a barrel and of the motion of the screws, it is reasonable to model a twin-screw extruder by this type of transfer. The expression of the Nusselt number $Nu_{Skelland}$ is based upon a structure identical with that of Todd:[21]

$$Nu_{Skelland} = 4.9\left(\frac{D_{ext} \cdot v \cdot \rho}{\mu}\right)^{0.57} \cdot \left(\frac{C_p \cdot \mu}{\lambda}\right)^{0.47} \cdot \left(\frac{D_{ext} \cdot N}{v}\right)^{0.17} \cdot \left(\frac{D_{ext}}{L}\right)^{0.37} = \frac{h_{m/b} \cdot D_{ext}}{\lambda} \quad (4.4)$$

where v is speed of material inside barrel ($v = N.B$ (m.s^{-1}), B is pitch of the screw (m) and L is length of the heat exchanger (m).

These two correlations show that it is necessary to know the material's physical and thermal characteristics (density, ρ (kg.m^{-3}), specific heat, C_p (J.kg^{-1}.K^{-1}) and thermal conductivity, λ (W.m^{-1}.K^{-1})) and rheological characteristics (dynamic viscosity of mixture, μ (Pa.s), dynamic viscosity on the wall of the barrel, μ_w (Pa.s).

(c) Transfer area between material and barrel

The power transferred between the material and the barrel is proportional to the transfer area. In the case of a twin-screw extruder, this area is proportional to the degree of fill of the screws defined by the operational conditions (rotation speed of the screws, pitch of the screws, material throughput, density of the material). The transfer area is the product of the internal surface area of the barrel and the degree of fill of the screws. The different expressions given in the literature are based upon the same structure. This is the ratio of the actual volume of material to the available volume. Mention may be made of the Booy[23] relationship which is the most routinely used for the solid and molten transport zones:

$$Tr = \frac{Q}{\rho \cdot N \cdot B \cdot S_{free}} \quad (4.5)$$

where Tr is degree of fill of screws (dimensionless), Q is throughput of material (kg.s^{-1}) and S_{free} available section between screws and barrel (m^2).

(d) Viscous dissipation: shear rate – viscosity – volume

Viscous dissipation is responsible for the conversion and intimate mixing of the material in the extruder. This thermal power of mechanical origin is a significant factor in the thermal changes in the material. These are generated by a velocity gradient, known as the shear rate (γ (s^{-1})), within a volume of material ($V_{sheared}$ (m^3)). Martelli[24] determines a mean shear rate in terms of a volume and calculates the viscous dissipations ($q_{shearing}$ (W)) such as:

$$q_{shearing} = \mu \cdot \gamma^2 \cdot V_{sheared} \quad (4.6)$$

He breaks down the total free space in the screws into four zones (Fig. 4.1) and considers a mean shear rate in each of these spaces. The total shear power is the sum of the powers in each of the four zones. Table 4.3 summarises the expressions given by Martelli.[24]

$$D_e = \frac{2}{\pi} \cdot (\pi \cdot D_{ext} - \sqrt{2 \cdot D_{ext} \cdot h} \quad (4.7)$$

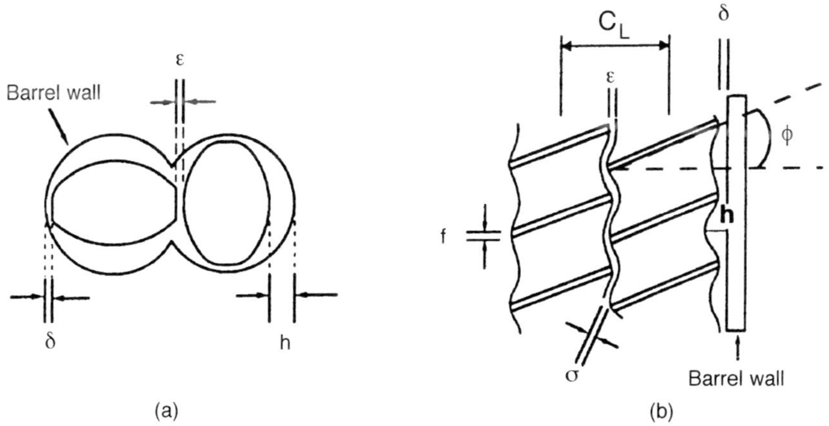

Fig. 4.1 Description of the different clearance between screws and barrel to illustrate the four zones defined by Martelli[24]: (a) front view, (b) axial view.

Table 4.3 Shear rate and volume in co-rotating twin-screw extruder according to Martelli[24] formulations

Zone	Shear rate	Volume
Channel: h	$(\pi \cdot D_e \cdot N)/h$	$n \cdot \frac{\pi^2}{2} \cdot D_e \cdot D_{ext} \cdot h \cdot \tan(\phi)$
Clearance between screw tips and barrel wall: δ	$(\pi \cdot D_{ext} \cdot N)/\delta$	$n \cdot f \cdot C_e \cdot \delta$
Clearance between screw tips and channel bottom: ε	$(\pi \cdot 2 \cdot C_L \cdot N)/\varepsilon$	$2 \cdot n \cdot f \cdot C_L \cdot \varepsilon$
Clearance flanks: σ	$(\pi \cdot C_L \cdot N)/\sigma$	$\frac{n}{2} \cdot \sqrt{(D_{ext}^2 - C_L^2)} \cdot h \cdot \sigma$

where $h, \phi, \delta, f, \varepsilon, C_L$ and σ are as shown in Fig. 4.1, D_e = equivalent diameter as in Equation (4.7) (m), n = number of flight of the screw (dimensionless), C_e = equivalent circumference = $\pi.D_e$ (m).

In order to quantify the shear rate more precisely it is necessary to determine the velocity profile in the volume in question. For this purpose, a study of flow through the screws is essential. In view of the complexity of the flow, simplifying assumptions must be made. Approaches in one, two or three dimensions developed by Tayeb,[25] Barrès[26] and Noé[27] based upon the solution of the Navier-Stokes equations for an isothermal, incompressible fluid behaving in a Newtonian manner, have been used to evaluate the velocity field. Also, finite element modelling of the flow in the screws by Szydlowski *et al.*,[28] Szydlowski and White,[29] White and Szydlowski,[30] and White and Chen[31] has provided further information about the velocity and pressure field.

(e) Difficulties in evaluating certain thermal parameters
This inventory of thermal parameters involved in extrusion cooking shows the limitations affecting their understanding and determination. The lack of knowledge of these parameters stems essentially from the difficulty of measuring them. The extreme conditions of temperature, shear, flow and pressure inside an extruder are difficult to reproduce using existing measurement facilities. If thermal calculations are to be valid, good estimates of all the parameters involved are necessary.

4.3 Experimental analysis

4.3.1 Introduction

If heat transfer in extrusion cooking is to be understood and controlled, it is essential to grasp certain physical quantities in the process. Experimental investigations make it possible to measure the parameters necessary for controlling a process, to deduce thermal quantities from the measurements, and to provide data for the physical models used to predict the thermal and mechanical behaviour of the material. The experimental approach must be both global and local. The parameters to be measured are:

- the energies involved: mechanical, heating, cooling and losses
- the temperature field in a barrel
- the profiles of material temperature and pressure along the extruder.

Special instrumentation will have to be developed to meet all these requirements. There has so far been little experimental work on heat transfer in an extruder. The research reported in the literature is based on limited measuring facilities and determined only the temperature and pressure in a localised fashion (in the die for Yacu,[13] and at several points along the extruder for Cardenas-Caroti *et al.*,[32] Noé[27], Mohamed[33] and Van Zuilichem *et al.*[34]). The temperature field in a barrel has never been measured in order to estimate the heat flow patterns. This information is, of course, fundamental because the thermomechanical processing of the material reflects the temperature of the barrel. All the thermal models are based on the assumption that the barrel is isothermal. The following paragraph describes an original experimental approach applied to a co-rotating twin-screw extruder. A detailed description of the experimental set-up is given and some of the more significant results are quoted.

4.3.2 Experimental set-up developed in recent years

Although a great deal of experimental work has been done on extrusion, very few extruders have been fully instrumented in order to monitor the thermal and mechanical changes in the material. It is worth noting that methods for measuring the material temperature at the end of the screw have been developed

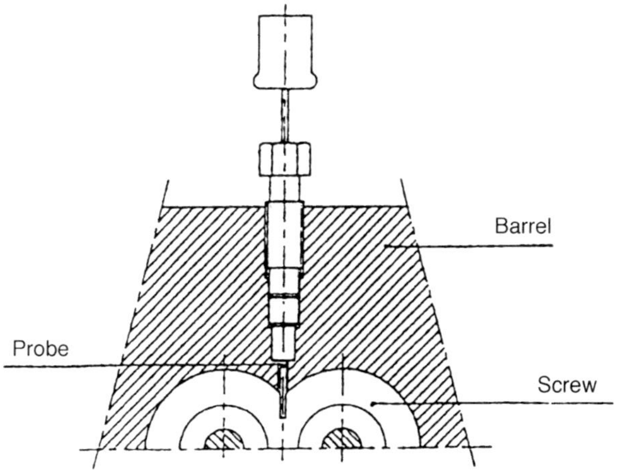

Fig. 4.2 Experimental set-up used by Van Zuilichem *et al.*[34] to measure the temperature of the material inside a twin-screw extruder.

for single-screw extrusion (Van Leeuwen *et al.*[35–37] and Schläffer *et al.*[38]). The experimental arrangements for investigating heat transfer in twin-screw extruders have been listed.

Van Zuilichem *et al.*[34] measure the material temperature along a twin-screw extruder using the experimental set-up shown in Fig. 4.2. They estimate that the thermocouples penetrate 7 mm into the product and that the temperature measured is in fact that of the product. However, it is not stated whether the thermocouples are inserted from the top or bottom of the barrel. Since the degree of fill of the screws is not a maximum along the whole length of the extruder, it is sensible to insert the thermocouples from the bottom of the barrel to ensure that they are in fact in contact with the product. This kind of precaution is essential if the temperature measurements are to be truly representative.

Cardenas-Caroti *et al.*[32] give details of the instrumentation of an industrial twin-screw extruder – Clextral BC45 – used by INRA and CEMEF (Fig. 4.3). This experimental rig was used for the work of Tayeb,[25] and Tayeb *et al.*[14] It is used to measure the material temperature and pressure in the shear zone consisting of a thread and a grooved reverse thread.

4.3.3 Strategy and methods of measurement in extrusion processing

A pilot barrel from a Clextral BC45 extruder was instrumented in order to investigate the heat transfer and the material pressure profiles. The temperature field in the barrel is measured by means of 100 thermocouples. The distribution of these (Figs 4.4 and 4.5) was selected on the basis of a screw profile applying very high thermomechanical stresses (positive screw: pitch = 25 mm and length = 100 mm; reverse screw: pitch = – 15 mm and length = 100 mm) in order to bring out the three-dimensional nature of the temperature field. The

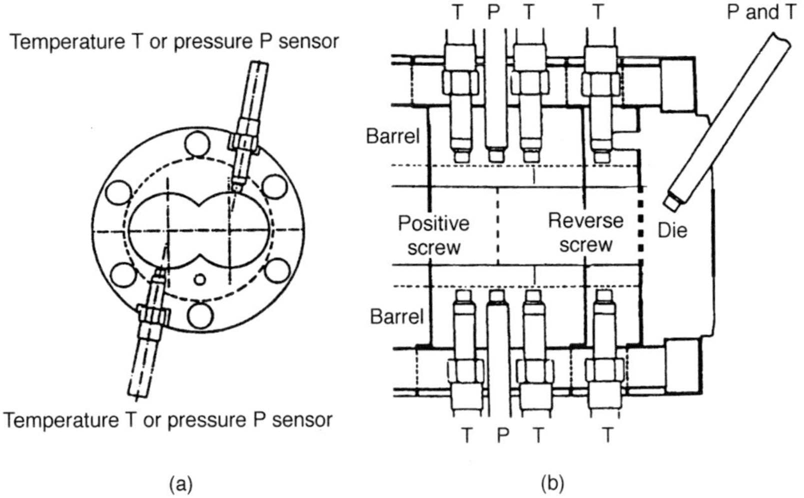

Fig. 4.3 Industrial twin-screw extruder Clextral BC45 instrumented by Cardenas-Caroti *et al.*[32]: (a) front view, (b) axial view.

thermocouples closest to the wall of the twin-screw circuit are 5 mm from it. The radial distance between two thermocouples is 6 mm. The angle between zones I and II is 62°. The angle between zones II and III is 56°. Zones IV, V and VI are obtained by symmetry with zones I, II and III. The axial variation of the material temperature is measured by four thermocouples positioned on the wall of the twin-screw circuit (Fig. 4.4 and 4.5). Figure 4.6 shows the instrumented barrel with its temperature sensors. Three pressure sensors are used to indicate the axial pressure profile in the material (Fig. 4.7). The thermocouples used are of the chromel-alumel type (K) 0.5 mm in diameter (barrel temperature) and 1.5 mm in diameter (material temperature). The pressure sensors are of the Dynisco TPT412 and TPT432A types covering a range from 0 to 1000 bar.

The mechanical power supplied by the motor is determined by measuring its speed of rotation (tachometer) and its torque (torquemeter). The heating power is determined by measuring the electric power consumed by the heater resistors. The cooling power is calculated from the energy balance of the air conditioning fluid (temperature difference and flow rate).

4.3.4 Results

The results given were obtained in the following operating conditions: type BC45 extruder with an L/D ratio of 18, maize meal, throughput 50 to 250 kg.h^{-1}, screw rotation speed from 150 to 500 rpm, moisture content (wet base) between 15 and 18%, and setpoint temperature from 140 to 170°C.

The temperature field in the barrel is illustrated by a typical axial profile (Fig. 4.8). This shows considerable non-uniformity of the temperatures in the barrel. Axial and radial temperature differences of 50°C and 20°C prove that there is

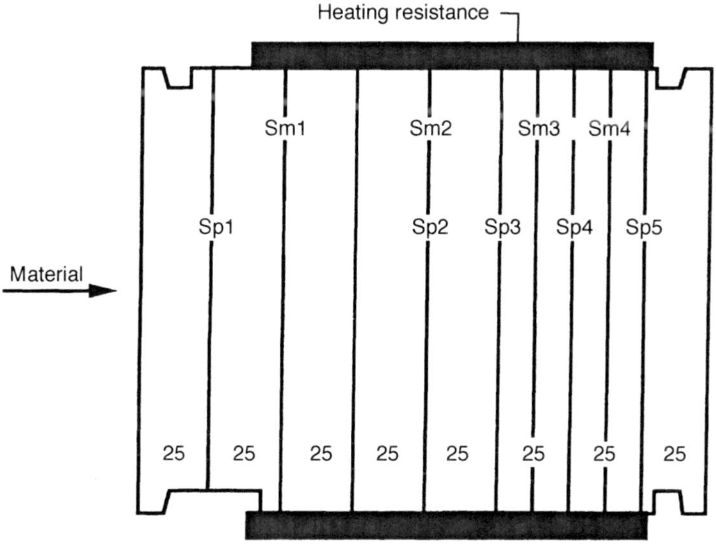

Fig. 4.4 Axial distribution of the different sections of thermocouples within the pilot barrel.

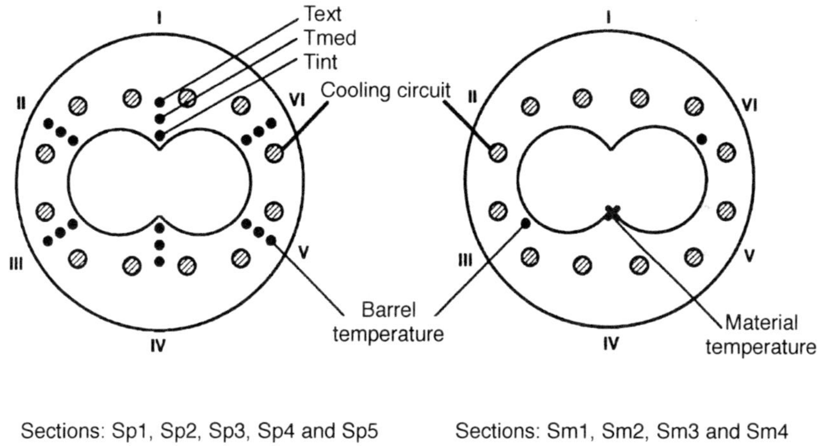

Fig. 4.5 Distribution of the thermocouples in all experimental sections within the pilot barrel.

heat transfer within the barrel. The assumption of an isothermal barrel in the thermal models is unrealistic. For better modelling of heat transfer, this temperature field must be taken into account.

Figure 4.9 illustrates the axial material temperature profile for different moisture contents (screw rotation speed, material throughput and setpoint temperature are 330 rpm, 150 kg/h and 150°C, respectively). Reducing the moisture content raises the material temperature in the reverse thread section

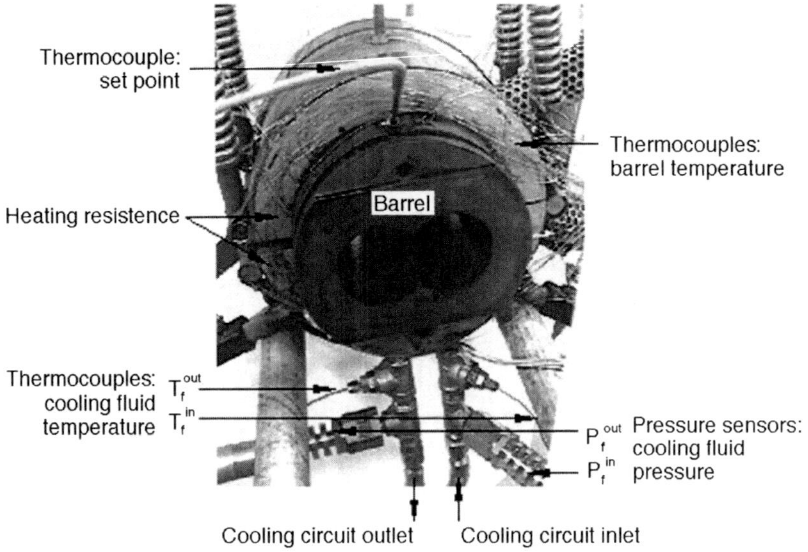

Fig. 4.6 Instrumented barrel of twin-screw extruder Clextral BC45 by Mottaz.[39]

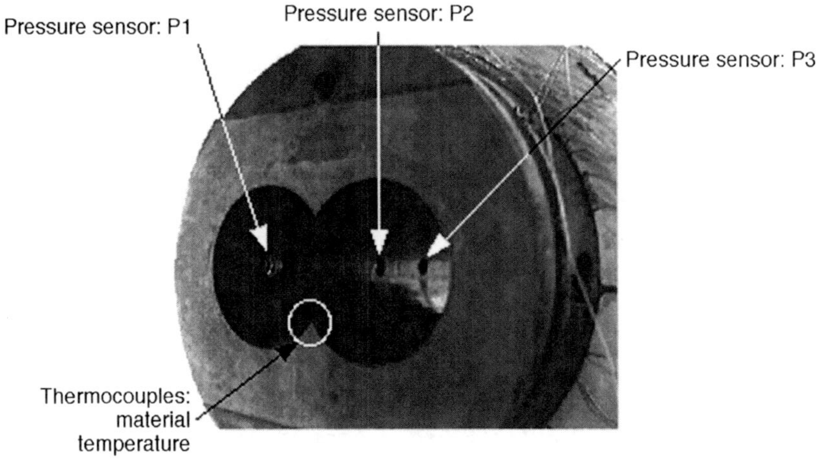

Fig. 4.7 Axial positions of pressure sensors and thermocouples within the pilot barrel.

only. A reduction of 1% in moisture content causes an increase of 8 to 10ºC in material temperature. Reducing the moisture content will also increase the viscosity of the maize meal. This change in viscosity brings about greater viscous dissipation and hence a higher material temperature.

Figure 4.10 shows how the pressures vary with time along the barrel. The pressure P1 measured in the thread does not exceed 2 bar. The pressure P2 measured at the transition between the thread and the reverse thread is very high and shows substantial fluctuations (30 to 90 bar) about a mean value (between

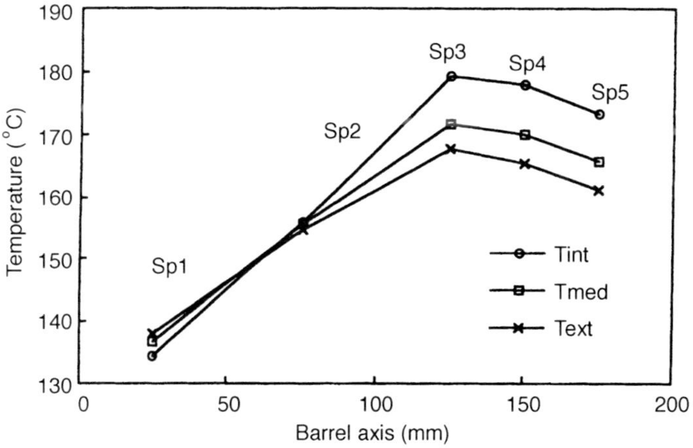

Fig. 4.8 Axial temperature profile inside the pilot barrel in zone III with operating conditions such as: maize flour, N = 300 rpm, Q = 200 kg.h^{-1}, MC (wet base) = 17% and setpoint temperature = 150°C.

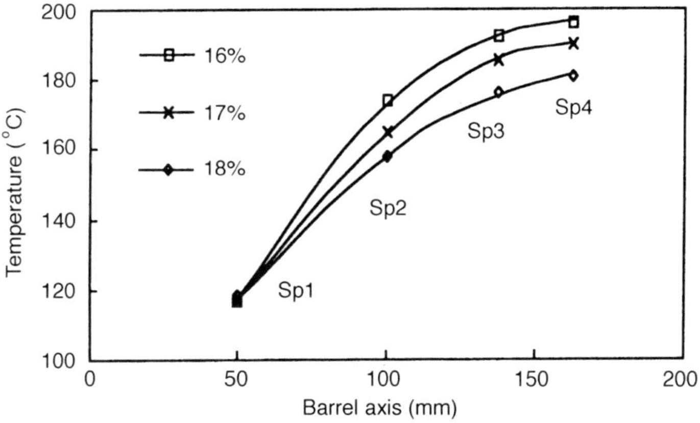

Fig. 4.9 Axial material temperature profile for different moisture content with operating conditions such as: maize flour, N = 300 rpm, Q = 150 kg.h^{-1} and setpoint temperature = 150°C.

50 and 200 bar). The mean value of the pressure P3 is between 18 and 45 bar, and the fluctuations are only a few bar.

It is interesting to compare the axial profiles of pressure and temperature in the material. The values of temperature and pressure taken together tell us about the behaviour of the material in the extruder. Figure 4.11 shows the axial profiles of the material temperature and pressure in the steady state. The increase in pressure at the transition from the thread to the reverse thread is accompanied by a considerable increase in material temperature in the thread. The material is transformed by the shear stresses generated by the reverse thread which cause

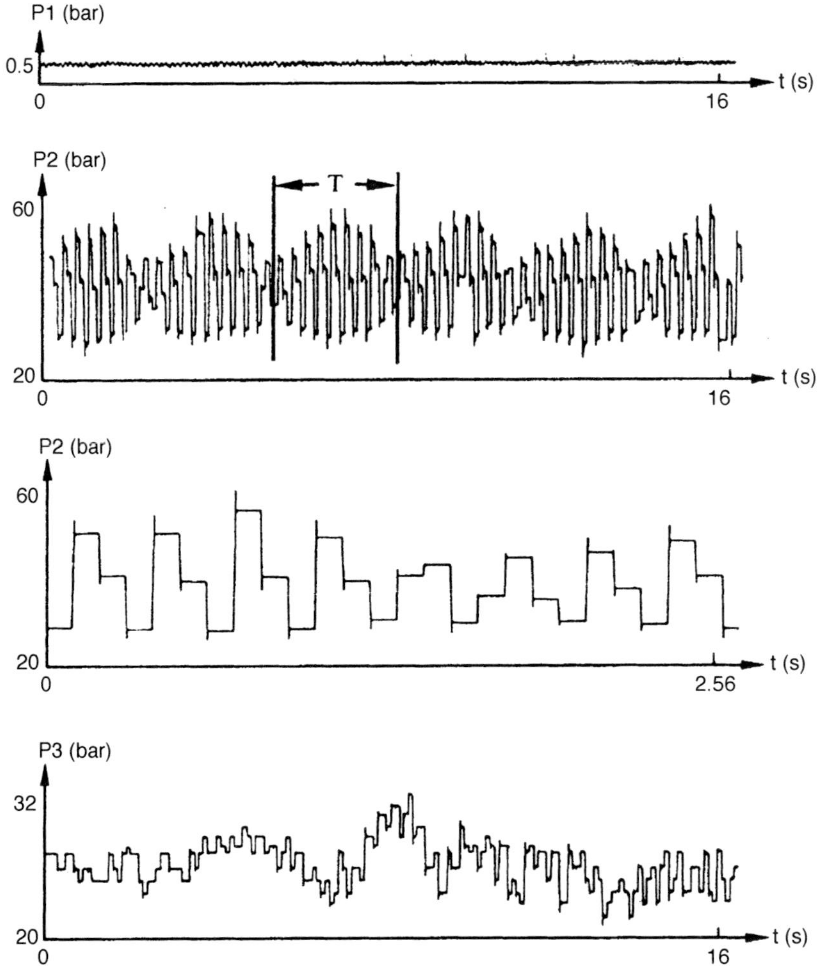

Fig. 4.10 Material pressure vs. time inside the pilot barrel with operating conditions such as: maize flour, N = 400 rpm, Q = 80 kg.h^{-1} and MC (wet base) = 17%.

the rise in temperature. This demonstrates the very close coupling between pressure and temperature in the conversion of the material.

4.3.5 References in the literature to research on heat transfer

Mosso *et al.*[40] made a systematic study of the influence of the operating conditions on the temperature and pressure in the die using a complex foodstuff (a mixture of several flours). The conclusions of their investigation are summarised in Table 4.4. They show that increasing the moisture content (at constant throughput and speed) reduces temperature and pressure. Increasing the speed (for constant throughput and moisture content) causes a slight rise in

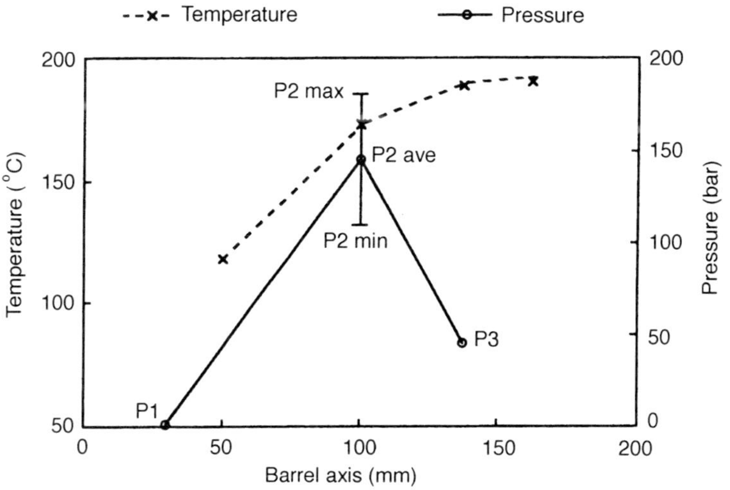

Fig. 4.11 Axial profiles of material temperature and pressure in steady state with operating conditions such as: maize flour, N = 400 rpm, Q = 200 kg.h^{-1} MC (wet base) = 17%.

Table 4.4 Effect of the operating conditions on the material temperature and pressure in the die (Mosso *et al.*[40])

Parameter	Operating conditions	Temperature (°C)	Pressure (bar)
$9 < MC < 20$	Q = 39 and N = 40	177 to 170	190 to 60
$9 < MC < 20$	Q = 39 and N = 100	185 to 170	100 to 40
$40 < N < 100$	Q = 40 and N = 15	172 to 177	110 to 90
$25 < Q < 50$	N = 97 and MC = 12	185 to 178	70 to 110
$40 < N < 100$ and $20 < Q < 50$	MC = 13 and Tr = 0.53	180 to 177	50 to 100

MC: moisture content (%), N: screw speed (rpm), Q: Flow rate of material (kg.h^{-1}), Tr: Filling ratio of the screw.

temperature and a more marked fall in pressure. The influence of the throughput (for constant speed and moisture content) is characterised by an increase in pressure and a fall in temperature. Changing the speed and throughput simultaneously for constant degree of fill causes an increase in pressure but does not affect the temperature. Fletcher *et al.*[41] investigate the extrusion cooking of maize flour for typical operating conditions ($Q = 30$ kg.h^{-1}, $N = 300$ and 400 rpm, and $MC = 25$ to 83% (wet base)). They measure the mechanical power, the temperature and pressure in the material along the extruder, and the characteristics of the product as it leaves the die. They show that the presence of the die causes a sudden increase in pressure and in temperature. The material temperature is between 75 and 150°C. The pressure in the die varies from 15 to 50 bar. The measurements made for different moisture contents show that the temperature and pressure fall when the moisture content rises. Rapid measurements of the pressure revealed fluctuations of 2 to 8 bar about a mean value.

Della Valle *et al.*[42,43] investigate the extrusion cooking of maize flour in typical operating conditions ($Q = 30$ kg.h^{-1}; $N = 210$ rpm; $MC = 21\%$ (wet base)). Increasing the moisture content causes the temperature and pressure to fall. Increasing the throughput at constant speed brings about a slight fall in temperature and an increase in pressure. The effect of the speed was not clearly established.

Although the research by Noé[27,44] concerned the extrusion of polypropylene, it does provide information about the effect of the operational conditions on the material pressure. The pressure at the transition from the thread to the reverse thread changed from 100 to 220 bar with fluctuations of about 50 bar for a throughput varying from 12 to 37 kg.h^{-1}, and a constant speed of 250 rpm. In the same conditions of throughput and speed, the pressure measured between the reverse thread and the die varied from 20 to 70 bar. The pressure in the die is between 40 and 80 bar, and the fluctuations are less marked (about 15 bar). This shows that increasing speed at constant throughput causes a fall in pressure, and that increasing throughput at constant speed increases the pressure.

4.3.6 Conclusions

By developing an experimental set-up dedicated to investigating heat transfer, we were able to formalise the thermal and mechanical behaviour of a barrel in the presence of maize meal in typical operating conditions. We have shown that the temperature field in the barrel shows axial, radial and angular non-homogeneities that induce conductive heat transfer within the barrel. The temperature fields in the barrel and in the material directly reflect the profile of the screw. The reverse thread generates a zone of very high compression at the thread–reverse thread transition. This compression of the material causes it to heat up intensely in the thread. Expansion takes place in the reverse thread where the temperature of the material reaches its highest value. The material is transformed under the combined effect of very high pressures and temperatures. Rapid acquisition of the pressure signal reveals a pulsation phenomenon at the thread–reverse thread transition. The presence of such pulsations suggests that the flow of the material in the vicinity of the transition zone is highly 'turbulent'. The effect of the operating conditions on temperature and pressure in the steady state is clearly established. The temperature rises if the speed increases or if the moisture content falls, and is reduced if the throughput increases. The pressure falls if the speed increases, and rises if the throughput increases.

So far, all the experimental work has been done in intermediate operating conditions (maximum values of throughput and speed of 250 kg.h^{-1} and 500 rpm respectively). The increasing capacity of extruders in terms of throughput (several hundred kg.h^{-1}) and speed (1200 rpm) may suggest changes in thermomechanical behaviour.

4.4 Thermal modelling

4.4.1 Introduction

In order to monitor and control the extrusion process, understanding and control of heat transfer are essential. Put another way, it is fundamental to determine the thermal behaviour of an extruder in order to obtain the required qualities of the transformed product. Accordingly it is necessary to develop appropriate measurement systems and to be capable of predicting the thermal performance of the extruder in given operating conditions. The parameters it is important to master are the temperature of the barrel, the temperature and pressure of the material, and the various powers involved. The models developed may be divided into two categories. The former are so-called 'knowledge' models that describe the physical phenomena involved by solving the conservation equations. The latter are so-called 'black box' representational models that provide simple relationships between the measured quantities and the operating parameters.

A different approach was devised for understanding and predicting heat transfer in an extruder barrel. A detailed physical model was developed using the finite element method. For industrial purposes, another, simplified, model was programmed for predicting the thermal behaviour of the barrel and the material in real-time.

4.4.2 Review of the best thermal models of extrusion

The main models for heat transfer in a twin-screw extruder are reviewed and compared. Only the most relevant 'knowledge' models will be considered, since the 'representational' models are not applicable whatever the operating conditions. The 'knowledge' models can be subdivided into three main topics:

- the description of flow: Tadmor and Klein,[45] Janssen,[46] Colonna *et al.*,[2] Tayeb *et al.*,[47]
- the energy aspects of the material and the extruder: Jepson,[48] Yacu,[13] Mohamed and Ofoli,[49] Tayeb *et al.*,[14] Chang and Halek,[16] Van Zuilichem *et al.*,[50] Della Valle *et al.*,[51]
- the rheological behaviour of the material: Harper *et al.*,[7] Fletcher *et al.*,[8] Vergnes and Villemaire,[9] Morgan *et al.*,[10] Mohamed *et al.*,[11] Vergnes *et al.*[12]

These models encompass the geometrical parameters of the screws and barrels, the operating conditions (material throughput, screw rotation speed and barrel setpoint temperatures), and the physical and thermal and rheological characteristics of the product or products. Considerable simplifying assumptions have to be made owing to the complexity of the geometry and the flow patterns. All these models rely upon solving the energy balance equation, the following common assumptions being made: steady state conditions apply, transfers are mono-dimensional, the barrel is isothermal, and the extruder can be broken down into functional zones according to Colonna *et al.*[2] The general expression of the balance equation is as follows:

$$Q \cdot C_p \cdot dT_m = h_{m/b} \cdot A \cdot dz \cdot (T_b - T_m) + dq_{shearing} \quad (4.8)$$

where A is circumference of the barrel wall (m) and dz is axial increment (m).

Each model differs in its expression of the heat transfer coefficient between the material and the barrel, and for viscous dissipation. We may quote the work of Yacu,[13] Della Valle *et al.*,[51] Potente *et al.*[52,53] and Vergnes.[54] They take into account the extruder as a whole and evaluate a large number of parameters such as the temperature, pressure, degree of fill, shear energy and the distribution of residence times.

Yacu[13] was the first to propose an overall thermomechanical model for extrusion cooking. In the zone of solid transport, viscous dissipation is assumed to be zero owing to partial filling of the screws. The convective heat transfer coefficient between material and barrel is set at 30 $W.m^{-2}.K^{-1}$. In the zone of pressurisation with change of phase and the shear zone (grooved reverse thread section), the transformation of the material is assumed to take place over a very short distance and the screws are considered to be full. The convective heat transfer coefficient between raw material and barrel is set at 500 $W.m^{-2}.K^{-1}$. According to the modelling of Martelli,[24] calculations show substantial viscous dissipation in this area.

Tayeb *et al.*[14] propose a theoretical and experimental approach to the extrusion cooking of maize starch. In the solid transport zone, there is a new description of product distribution based upon experimental observations (stopping/opening the extruder). They introduce a value for the power generated by friction of the material in the space between the top of the thread and the barrel wall to obtain temperatures at the end of the transport zone close to the experimental values (about 170C). The heat transfer coefficient varies from 400 to 2000 $W.m^{-2}.K^{-1}$. Barrès *et al.*[15] propose an alternative to this modelling approach such that the temperature of the material at the end of the transport zone is below its melting point (in accordance with the initial assumption). In the zone of pressurisation in the molten state, the Navier-Stokes equations for an incompressible isothermal fluid behaving in a Newtonian manner are solved in order to determine the pressure gradient and the velocity profile. The shear zone, established by a grooved reverse thread section, was particularly investigated by Tayeb *et al.*[47]

Overall thermomechanical modelling of the twin-screw extrusion process is possible using the LUDOVIC® program (Logiciel pour l'Utilisation des Doubles Vis Corotatives) (Program for the use of co-rotating twin-screw extruders). This program is a joint product of CEMEF and INRA. We may mention the dissertations of Tayeb,[25] Barrès[26] and Noé[44] together with the various associated publications: Della Valle *et al.*,[51] Tayeb *et al.*,[55,56] Vergnes,[54] Tayeb *et al.*,[14] Barrès *et al.*,[15] Vergnes *et al.*,[57] Delamare.[58]

We may mention other thermomechanical models involving functional zones that are variants of the above research. Mohamed and Ofoli[59] experimentally determined the barrel/material heat transfer coefficient (between 191 and 768 $W.m^{-2}.K^{-1}$) and the shear rate. Chang and Halek[16] propose a method for calculating the friction power in the solid state transport zone and the viscous dissipation in kneading systems. Van Zuilichem *et al.*[34, 50, 60–62] propose a thermomechanical model involving a method for calculating the material/barrel heat transfer coefficient using a correlation devised by Jepson.[48] This aspect of modelling lacks precision and is difficult to apply.

4.4.3 A new modelling approach

(a) Objectives and methods

An experimental approach to heat transfer in extrusion cooking is necessary but insufficient for predicting thermal changes in the barrel and the material, particularly when the operating conditions change. A heat transfer model has been developed for predicting the thermal behaviour of the barrel and the material in real-time. Such a model is developed in two stages:

- A physical model of heat transfer in a barrel is developed using a computer code involving the finite element method. The aim of this stage is to quantify the thermal stresses affecting the barrel. This detailed model (denoted DM) is calibrated and validated by simple experiments.
- The second stage is to reduce the detailed model into one providing a real-time response at a limited number of points (denoted RM – 'reduced model'). For this purpose a method for reducing thermal models by identification developed by Petit[63] was applied. This method was developed on the basis of automatic control theory and is precisely suited for the control-command of a process.

(b) Finite element model: temperature field inside barrel – thermal boundaries

The development of a model with a computer code using the finite element method makes it possible to calculate the temperature field T_b at all points on the barrel. Such a method is based upon solving the heat transfer equation in non-steady conditions:

$$\rho_b \cdot C_{p_b} \cdot \frac{\partial T_b}{\partial t} = \lambda_b \cdot \Delta T_b + \phi \tag{4.9}$$

where ρ_b is density of barrel (kg.m^{-3}), C_{p_b} is specific heat of barrel (J.kg^{-1}.K^{-1}), t is time (s), λ_b is thermal conductivity of barrel (W.m^{-1}.K^{-1}), Δ is a Laplacian operator and ϕ thermal stresses (W).

The aim of the detailed model is to identify the thermal stresses affecting the barrel:

- the heat input from the heater collar: $\phi_{heating}$
- losses to the exterior: ϕ_{losses}
- the source term generated by the shearing of the material: $\phi_{material}$
- the heat absorbed by the cooling liquid: $\phi_{cooling}$

These thermal stresses are the 'inputs' to the thermal model. Its 'outputs' are the changes with time of the temperature at various points on the barrel: $T_b\,(x, y, z, t)$.

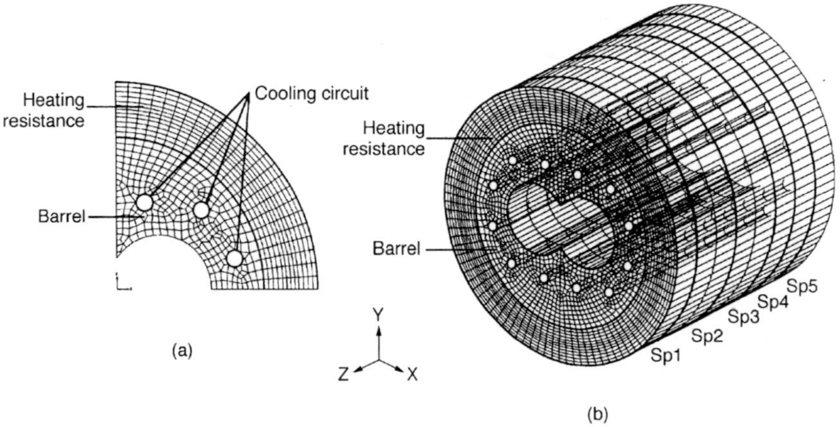

Fig. 4.12 Mesh used for the detailed model DM: (a) two-dimensional, (b) three-dimensional.

The thermal inputs to the model are identified independently of one another. The barrel geometry has symmetries that permit two-dimensional and three-dimensional representations for the detailed model (Fig. 4.12). The simplest mesh has 600 nodes for 700 elements while the most complicated has 23,000 nodes for 26,000 elements. A thermal model using finite elements allows the temperature field at all points on the barrel to be calculated with great refinement. The thermal stresses are identified by comparing the calculated and measured temperature fields. However, a computer code of this kind requires substantial data processing facilities (both hardware and software) and the computing times are long (from a few minutes to several hours). This does not lend itself to industrial use. A model must be inexpensive in both hardware and in time, and the overall behaviour must be available from a few representative points.

(c) Model reduction: real-time predictive model

A method of thermal model reduction using identification can reduce computing time as well as the number of points necessary for describing the thermal behaviour of the barrel. This model reduction method was developed and validated by Petit[63] and various researchers have worked on it (Petit *et al.*,[64] Hachette,[65] Petit *et al.*,[66] Petit and Hachette[67]). Model reduction is applied to systems that meet the assumptions of invariance, linearity and reciprocity. By successive conversions of the diffusion equation (discretisation, modal representation, and identification of specific modes) the identification structure of the reduced model is illustrated by Fig. 4.13. Reducing the model involves minimising a quadratic criterion J such that $y(t)_{RM} \approx y(t)_{DM}$ using an appropriate algorithm for defining the matrices F_r, G_r, H_r and K_r associated with each input $u(t)$. The quadratic criterion J to be minimised is defined as the square of the difference between the outputs $y(t)_{DM}$ and $y(t)_{RM}$ summed for each outlet and for each time step, as shown in:

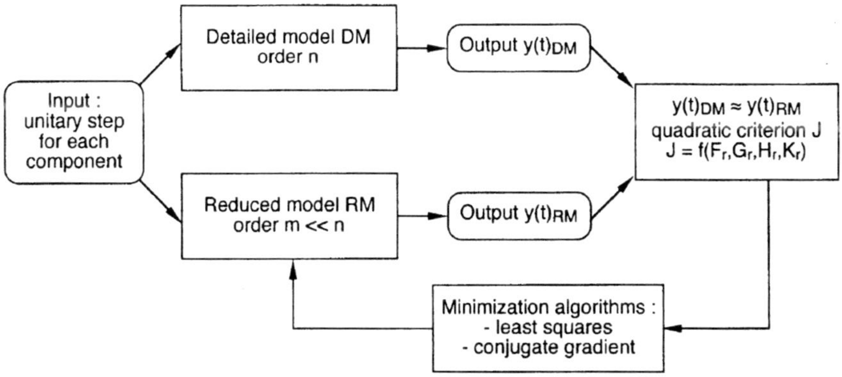

Fig. 4.13 Identification structure of the reduced model RM.

$$J(m, F_r, H_r) = \sum_{i=1}^{q} \sum_{k=1}^{r} (y_i(\tau_k)_{DM} - y_i(\tau_k)_{RM})^2 \tag{4.10}$$

where τ_k discretization moment (s), r = number of points of temporal discretization, i = index representating the *i*th component of the output vectors (dim. q).

Each system input is assigned a time constant so as to characterise its dynamic effect on the system outputs. These time constants are determined by relaxation thermograms of the temperatures at the desired points. It is then a matter of investigating the response of the system to inputs of the stepped type. The relaxation thermograms are obtained by switching the system from its initial condition to a steady state condition by activating a single one of the system inputs. From this steady state, the activated input is inhibited to allow the system to return to its initial condition. The temperature relaxation considered is approximated by a linear combination of the exponentials of the real time constants of the system. The mathematical formulation of the relaxation thermogram is then written:

$$T(t) = \sum_{j=1}^{m} \left\{ a_j \cdot \exp\left(\frac{-t}{\tau_j}\right) \right\} \tag{4.11}$$

where m (dimensionless), a_j (°C) and τ_j (s).

The vectors $[a_j, \tau_j]$ characterise the dynamic behaviour of the system with respect to each input. These time constants are used by the reduced model computer code for predicting thermal changes in the system.

(d) Comparison between the detailed model and the reduced model

A particular application to an extruder barrel is given to illustrate the comparison between the detailed and reduced models. The simulation involves heating the barrel using the heat flow resulting from the shearing of the material to reach a steady state, then generating alternate calls for heating and cooling

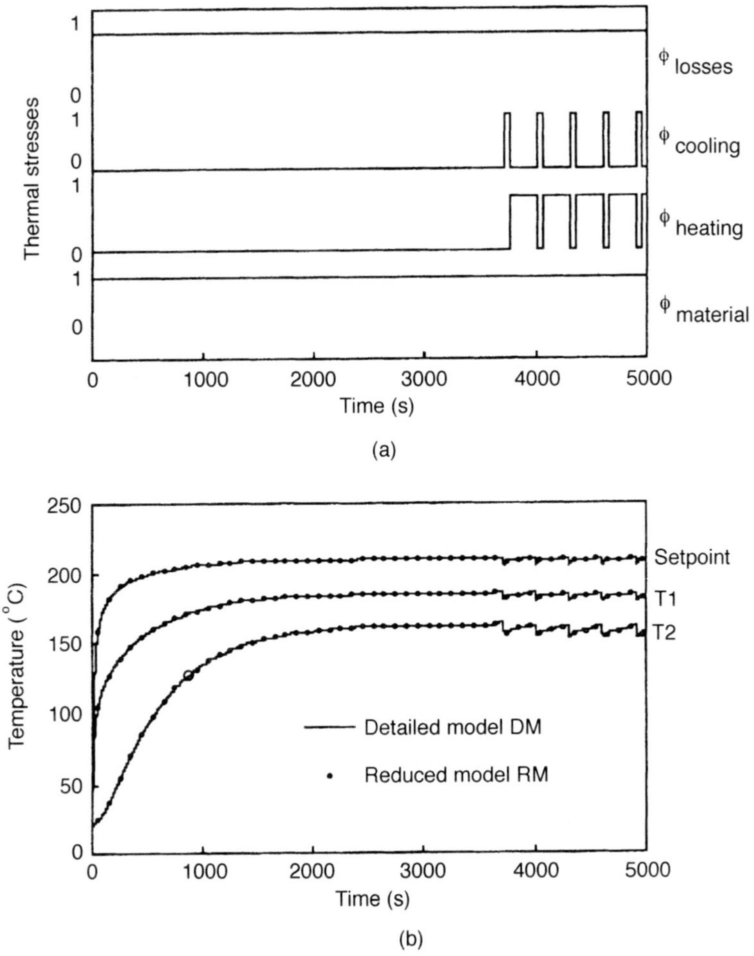

Fig. 4.14 Numerical simulations: (a) evolution vs. time of the thermal stresses affecting the barrel, (b) temperature vs. time for three outputs calculated with detailed and reduced models.

(simulating a control system). The changes over time of three nodes (setpoint, T1 and T2) for the detailed and reduced models are illustrated in Fig. 4.14. The computing time for a simulated duration of 5000 s is 15,300 s for the detailed model and 3 s for the reduced model.

4.4.4 Conclusions

Review of the existing thermal models for extrusion cooking shows that research has focused on the material at the expense of the barrel. Our measurements of the temperature field in the barrel have demonstrated the three-dimensional nature of the heat transfers. This justifies our choice of heat transfer modelling which takes into account the coupling between barrel and material. The results obtained with our reduced model confirm the validity of the method used.

4.5 Sizing an extruder and future trends

4.5.1 Introduction

Research carried out over the last 15 years illustrates the major role of heat transfer in the extrusion cooking process. It is important to be able to measure the physical quantities related to the thermomechanical treatment applied to the material. It is also valuable to be able to predict the thermal behaviour of an extruder with a view to its control and optimisation. The aim of this section is to propose a simple method for calculating the thermal performance of an extruder barrel. The limiting factor on heat transfer between the material and the barrel is clearly identified, and an experimental method for estimating the heat transfer coefficient between the material and the barrel presented, and details of the calculation of the cooling capacity of a barrel given. Ways and means of increasing the power transferred between the material and the cooling fluid are proposed. Existing and future processes are discussed to provide an overall view of thermal processes in extrusion cooking and of the subsequent operations carried out on the extruded material. Current developments tend to show that the thermal processes in extrusion cooking should not be regarded as the only physical phenomenon that transforms the material. The links between the thermal processes, mechanical processes and rheology must be investigated, including the interactions between these modes of transfer.

4.5.2 Sizing and optimisation of an extruder: objectives and methods

The objectives of sizing and thermally optimising an extruder are the homogenisation of the temperature field in the barrel and the control of the material temperature. The design of an extruder must incorporate the thermal components of the process. The various energies involved must be identified: viscous dissipations, cooling power and heating power. The items to be defined are the barrel geometry, the cooling circuit, the heating system and the location of the regulating sensor.

The heating and cooling actuators are sized so as to have short response times and to limit the temperature gradients in the barrel. The ratios between the powers and the geometrical dimensions of the barrel must be adhered to so as to take into account the differentials between the various heat transfer coefficients: raw material/barrel and cooling fluid/barrel. The starting point in the thermal design of a barrel is the estimate of the viscous dissipation that may be generated by the shearing of the material. In fact, a barrel will be optimised if its cooling circuit is capable of extracting this power so as to maintain a constant temperature in the barrel.

4.5.3 Improvement of temperature control in an extruder

(a) Power transferred and equivalent heat transfer coefficient

The heat transfers between the material and the cooling fluid involve convective and conductive processes. There is convective heat transfer between the material and the barrel (Equation 4.12), conductive transfer in the barrel (Equation 4.13), and convective heat transfer between the barrel and the fluid (Equation 4.14). On this basis the heat transferred between the material and the fluid can be expressed in different ways. The different expressions of the power exchanged, ϕ (W), are as follows:

$$\phi = h_{\mathrm{b/m}} \cdot S_{\mathrm{b/m}} \cdot (T_{\mathrm{m}} - T_{\mathrm{b}}^{\mathrm{m}}) \tag{4.12}$$

$$\phi = \frac{2\pi l}{\ln\left(\frac{R}{r}\right)} \cdot \lambda_b \cdot [T(R) - T(r)] \tag{4.13}$$

$$\phi = h_{\mathrm{f/b}} \cdot S_{\mathrm{f/b}}(T_{\mathrm{b}}^{\mathrm{f}} - T_{\mathrm{f}}) \tag{4.14}$$

where $T_{\mathrm{b}}^{\mathrm{m}}$ is temperature of barrel wall in contact with material (ºC), $h_{\mathrm{f/b}}$ is convective heat transfer coefficient between cooling fluid and barrel ($\mathrm{W.m^{-2}.K^{-1}}$), $S_{\mathrm{f/b}}$ is exchange area between cooling fluid and barrel ($\mathrm{m^2}$), $T_{\mathrm{b}}^{\mathrm{f}}$ is temperature of barrel wall in contact with cooling fluid (ºC), T_{f} is temperature of cooling fluid (ºC), R (m), r (m), l (m).

It is clear that to determine this heat flow, it is necessary to know the wall temperatures (material – $T_{\mathrm{b}}^{\mathrm{m}}$ – and fluid – $T_{\mathrm{b}}^{\mathrm{f}}$). Since it is impossible to measure these temperatures, the heat flow is calculated from the power absorbed by the fluid, on the assumption that all the power contributed by the shearing of the material is absorbed by the cooling fluid. A new expression of the heat flux is:

$$\phi = m_{\mathrm{f}} \cdot C_{\mathrm{P_f}} \cdot (T_{\mathrm{f}}^{\mathrm{out}} - T_{\mathrm{f}}^{\mathrm{in}}) \tag{4.15}$$

where m_{f} is cooling fluid flow rate ($\mathrm{kg.s^{-1}}$), $C_{\mathrm{P_f}}$ is specific heat of cooling fluid ($\mathrm{J.kg^{-1}.K^{-1}}$), $T_{\mathrm{f}}^{\mathrm{out}}$ and $T_{\mathrm{f}}^{\mathrm{in}}$ are, respectively, outlet and inlet temperature of cooling fluid (ºC).

There is another way of calculating the heat transfer. For this purpose the equivalent thermal resistance, $R_{\mathrm{th}}^{\mathrm{equi}}$ ($\mathrm{K.W^{-1}}$), of the system could be calculated as:

$$R_{\mathrm{th}}^{\mathrm{equi}} = R_{\mathrm{th}}^{\mathrm{fluid}} + R_{\mathrm{th}}^{\mathrm{barrel}} + R_{\mathrm{th}}^{\mathrm{material}} = \left(\frac{1}{h_{\mathrm{f/b}} \cdot S_{\mathrm{f/b}}}\right)^{\mathrm{fluid}} + \left(\frac{e}{\lambda_{\mathrm{b}} \cdot S_{\mathrm{ave}}}\right)^{\mathrm{barrel}} + \left(\frac{1}{h_{\mathrm{m/b}} \cdot S_{\mathrm{m/b}}}\right)^{\mathrm{material}} \tag{4.16}$$

where e is distance between cooling fluid and material (m) and S_{ave} is average area calculated as shown in Equation 4.17 ($\mathrm{m^2}$):

$$S_{\mathrm{ave}} = \frac{S_{\mathrm{f/b}} - S_{\mathrm{m/b}}}{\ln(S_{\mathrm{f/b}} - S_{\mathrm{m/b}})} \tag{4.17}$$

This equivalent thermal resistance is the sum (the transfers are in series) of the thermal resistances of the system (cooling fluid: R_{th}^{fluid}, barrel: R_{th}^{barrel} and material: $R_{th}^{material}$ (K.W^{-1})). This equation involves heat transfer coefficients and transfer areas. The thermal conductivity of the barrel is known (about 40 W.m^{-1}.K^{-1}). The heat transfer coefficient of the fluid is estimated from correlations. The transfer coefficient of the material is the term that is subject to most uncertainty.

(b) Limit on heat transfer: restrictive factor in heat transfer

The heat flux transferred is the product of the equivalent heat transfer coefficient, $(h.S)_{equi}$ (W.K^{-1}), and the logarithmic average temperature difference between material and cooling fluid, $\Delta MLT_{m/f}$ (°C):

$$\phi = (h.s)_{equi} \cdot \Delta MLT_{m/f} = \left(\frac{1}{R_{th}^{equi}}\right) \cdot \Delta MLT_{m/f} \tag{4.18}$$

If heat transfer is to be improved, these two parameters must be modified. For the equivalent heat transfer coefficient, the limiting element is that whose thermal resistance is highest. By comparing the thermal resistances of the material, the barrel and the fluid, it can be shown that the limitation on heat transfer comes from the material. In fact the material/barrel heat transfer coefficient is low compared with that between the fluid and the barrel. In normal practice the former is taken as 500 W.m^{-2}.K^{-1} while the latter is 5000 W.m^{-2}.K^{-1}.

(c) Convective heat transfer coefficient between cooling fluid and barrel

The convective heat transfer coefficient $h_{f/b}$ between the cooling fluid and the barrel is calculated using the Dittus-Boelter equation:

$$\mathrm{Nu_f} = 0.023 \cdot \mathrm{Re_f^{0.8}} \cdot \mathrm{Pr_f^{0.4}} = \frac{h_{f/b} \cdot d_f}{\lambda_f} \tag{4.19}$$

where $\mathrm{Nu_f}$ is Nusselt number (dimensionless), $\mathrm{Re_f}$ is Reynolds number (dimensionless) and $\mathrm{Pr_f}$ is Prandtl number (dimensionless) ($\mathrm{Re_f}$ and $\mathrm{Pr_f}$ are calculated for the cooling fluid).

This equation is valid for the forced convection inside a circular tube in turbulent conditions. Using this equation it can be shown that the heat flux transferred is proportional to the ratio of the cooling fluid flow rate, m_f (kg.s^{-1}), to the tube diameter, d_f (m):

$$\phi \propto \left(\frac{m_f}{d_f}\right)^{0.8} \tag{4.20}$$

It can be seen therefore that the smaller the diameter the higher the transfer coefficient, and hence the better the heat transfer (for a given fluid flow). The value of 5000 W.m^{-2}.K^{-1} corresponds to a fluid flow rate above which any increase in the flow produces only a very small increase in heat transfer.

The optimum diameter and flow rate cannot be determined only by calculating the heat transfer coefficient. It is also necessary to calculate the pressure drops in the circuit, because these increase as the diameter falls. It is therefore necessary to find a compromise between the heat transfer coefficient and the pressure drop inside cooling circuit, ΔP_f (mCE), calculated using the generic equation:

$$\Delta P_f = \left(\sum_{i=1}^{n} \zeta_i + \sum_{j=1}^{m} \frac{\lambda_j \cdot l_j}{d_j} \right) \cdot \left(\frac{U^2}{2 \cdot g} \right) \tag{4.21}$$

where n is number of specific losses (dimensionless), m is number of regular losses (dimensionless), U is velocity of the cooling fluid (m.s^{-1}) and g is gravity constant (m.s^{-2}).

These consist of regular losses – the straight length of the tube – ($\lambda_j \cdot l_j / d_j$) and specific losses – at elbows, restrictions, and so on – (ζ_i). A reference book for calculating pressure drops was prepared by Idel'cik.[68]

(d) Experimental method for measuring the convective heat transfer coefficient between foodstuff and barrel

A simple experimental method can be used for estimating the convective heat transfer coefficient between the material and the barrel. This involves measuring the variation in material temperature (input/output) and the power absorbed by the cooling fluid (temperature difference and flow rate). Knowing the amount of heat transferred, the equivalent heat transfer coefficient $(hS)_{equi}$ can be deduced using Equation 4.18. From the equivalent coefficient it is possible to determine the material/barrel heat transfer coefficient $h_{m/b}$ using Equation 4.16 (the cooling fluid/barrel heat transfer coefficient $h_{f/b}$ being determined with Equation 4.19 from knowledge of the flow rate).

(e) Improving the heat transfer between foodstuff and cooling fluid

Heat transfer can be improved by modifying the geometrical parameters and optimising the operating conditions of the extruder. The geometrical parameters are the diameter of the cooling circuit, the distance between the cooling circuit and the material, and the heat transfer areas (material and fluid). The operating conditions (screw rotation speed, throughput and screw profile) must be chosen to give the highest possible degree of fill in the screw. The amount of heat transferred is proportional to the transfer area. By these ways, the power transferred between material and cooling may be increased.

4.5.4 Future trends

By definition, extrusion cooking is a process of mechanical and thermal transformation. It is necessary to draw a distinction between routine applications (snacks, pet food, expanded cereals, etc.), for which the importance of heat is secondary, and applications necessitating enhanced thermal performance. New high added value processes have emerged as the thermal engineering of the extruder (barrels and die) has been developed. The most striking example is extrusion cooking in a moist medium (CEMH). This process of fibration of plant and animal proteins is developing rapidly today. Cheftel *et al.*[69] have devoted particular attention to this method of processing foodstuffs. The viability of this process stems from the control of the thermal attributes of the die. The reactive extrusion conversion of casein into caseinate (sodium or calcium caseinate) is another important example where the cooling capacity of the barrels must be a maximum. This material in fact requires intense cooling because its high viscosity leads to substantial viscous dissipation. Pellet processing also requires a high level of temperature control and cooling efficiency in the last section of the extruder; better quality of the product can thus be obtained as a result of heat engineering. These few examples show that heat engineering is a source of innovation as regards new products.

However, heat engineering in extrusion cooking is not limited to the extruder and the die. The operations that precede and follow the extruder must be incorporated. Upstream, many foodstuffs are precooked or premoistened before being injected into the extruder. This operation is done in a preconditioner and involves injecting steam into a flour so as to improve the efficiency of the extruder (better inputting). Mention may also be made of thermostatically-controlled injection tanks for feeding material into the extruder. Downstream, the extruded materials require further processing to obtain a finished product. The operations of cutting, shaping, coating and drying are an integral part of extrusion cooking. Future developments in extrusion cooking will involve the design of comprehensive lines for which skills in heat engineering, mechanical engineering, rheology and fluid mechanics will be necessary.

Technical innovations will be needed for improving the thermal and mechanical performance of extruders. The geometry of the screws will have to be modified to meet the increasing requirements regarding wear and replacement. The capacity of modern extruders (1200 rpm – 100 $kg.h^{-1}$ to several $t.h^{-1}$) suggests further developments in heat transfer.

4.6 Conclusions

The extrusion cooking conversion has developed substantially over the last twenty years as regards its flexibility and its potential in terms of products. The process involves complex phenomena of heat and mass transfer and momentum. The quality of the product leaving the die is determined by its thermomechanical history as it passes through the barrel of the extruder. The description of the different heat transfers given in section 4.2 clearly shows the importance of heat engineering in extrusion cooking. The physical parameters that play a role in the heat transfers have been clearly identified. Some of these parameters are difficult to estimate because of the severe environmental conditions in a barrel.

For these reasons, the study of heat transfer in extrusion cooking must be tackled using appropriate tools and methods, both in experiment and modelling. Considerable research has been done over the last decade into the thermo-mechanical aspects of the twin-screw extrusion process. We may mention the LUDOVIC® program for modelling of which a summary description has been given by Tayeb *et al.*[56] The small number of experimental set-ups used (Van Zuilichem *et al.*,[34] Cardenas-Caroti *et al.*[32]) have concentrated on the thermal behaviour of the material only. No author has taken an interest in the link between the temperature field in a barrel and the thermal evolution of the material.

An original methodology for both experiment and modelling has been developed. The wealth of instrumentation used on a barrel is a new departure in extrusion. The application of a new modelling method enables the thermal behaviour of the barrel to be linked to that of the material. Our results are relevant and show that the methods adopted are valid. We present a simplified experimental method for estimating the convective heat transfer coefficient between the material and the barrel.

Future research into the extrusion cooking process will certainly be devoted to correlate better the interaction between the converting material and the mechanics. The content of this chapter illustrates this tendency.

4.7 Acknowledgements

This study received financial and technical support from Clextral company. The authors wish to thank particulary Prof. L. Tadrist from Institut Universitaire des Systèmes Thermiques Industriels for his valuable contribution for this study. J. Mottaz obtained a PhD in Thermal Engineering in 1997.

4.8 References

1. ZOULALIAN A., 'Macromélange et micromélange – Cas de l'extrusion (13–19)', *40 ans d'extrusion bi-vis chez CLEXTRAL*, Firminy, Clextral, 1996.
2. COLONNA P., MELCION J.P., VERGNES B. and MERCIER C., 'Flow, mixing and residence time distribution of maize starch within a twin-screw extruder with a longitudinally-split barrel', *Journal of Cereal Science*, 1983 **1** 115–25.
3. DELLA VALLE G. and KOZLOWSKI A., 'Bilan énergétique d'un cuiseur-extrudeur bi-vis. Premiers résultats appliqués à l'amidon', *Cuisson-Extrusion Colloque No. 41*, Nantes, INRA Editions, 1986.
4. DELLA VALLE G., KOZLOWSKI A., COLONNA P. TAYEB J., 'Starch transformation estimated by the energy balance on a twin-screw extruder', *Lebensm. Wiss. u. Technol.*, 1989 **22** 279–86.
5. DELLA VALLE G. and VERGNES B., *La cuisson-extrusion: Propriétés thermophysiques et rhéologiques des substrats utilisés en cuisson-extrusion (440–467)*, Paris, Lavoisier Tec&Doc, 1994.

6. COLONNA P. MERCIER C., 'Gelatinisation and melting of maize and pea starches with normal and high amylose genotypes', *Phytochemistry*, 1985 **24** 8 1967.
7. HARPER J.M., RHODES T.P. WANNINGER L.A., 'Viscosity model for cooked cereal doughs', *AIChE Symposium Series*, 1971 **67** 40–3.
8. FLETCHER S.I., MCMASTER T.J., RICHMOND P. and SMITH A.C., 'Rheology and extrusion of maize grits', *Chem. Eng. Commun.*, 1985 **32** 239–62.
9. VERGNES B. and VILLEMAIRE J.P., 'Rhelogical behaviour of low moisture molten maize starch', *Rheologica Acta*, 1987 **26** 6 570–6.
10. MORGAN R.G., STEFFE J.F. and OFOLI R.Y., 'A generalized viscosity model for extrusion of protein doughs', *Journal of Food Process Engineering*, 1989 **11** 55–78.
11. MOHAMED I.O., OFOLI R.Y. MORGAN R.G., 'Modeling the average shear rate in a co-rotating twin-screw extruder', *Journal of Food Process Engineering*, 1990 **12** 227–46.
12. VERGNES B, DELLA VALLE G. and TAYEB J., 'A specific slit die rheometer for extruded starchy products. Design, validation and application to maize starch', *Rheologica Acta*, 1993 **32** 5 465–76.
13. YACU W.A., 'Modeling a twin-screw co-rotating extruder', *Journal of Food Engineering*, 1985 **8** 1–21.
14. TAYEB J., VERGNES B. and DELLA VALLE G., 'A basic model for a twin-screw extruder', *Journal of Food Science*, 1988 **53** 4 1047–56.
15. BARRÈS C., VERGNES B. and TAYEB J., 'An improved thermal model for the solid conveying section of a twin-screw extrusion cooker', *Journal of Food Engineering*, 1991 **15** 167–85.
16. CHANG K.L.B. and HALEK G.W., 'Analysis of shear and thermal history during co-rotating twin-screw extrusion', *Journal of Food Science*, 1991 **56** 2 518–31.
17. MOHAMED I.O. and OFOLI R.Y., 'Average heat transfer coefficients in twin-screw extruders', *Biotechnology Progress*, 1989 **5** 4 158–63.
18. LEVINE L. and ROCKWOOD J., 'A correlation for heat transfer coefficients in food extruders', *Biotechnology Progress*, June 1986 **2** 2 105–8.
19. MOHAMED I.O., MORGAN R.G. and OFOLI R.Y., 'Average convective heat transfer coefficients in single screw extrusion of non-Newtonian food materials', *Biotechnology Progress*, 1988 **4** 2 68–75.
20. LE ROUX D., VERGNES B., CHAURAND M. and ABECASSIS J., 'A thermomechanical approach to pasta extrusion', *Journal of Food Engineering*, 1995 **26** 3 351–68.
21. TODD D., 'Heat transfer in twin-screw extruders', *Society of Plastics Engineers ANTEC*, 1988 54–8.
22. SKELLAND A.H.P., 'Correlation of scraped-film heat transfer in the votator', *Chemical Engineering Science*, 1958 **7** 166–75.
23. BOOY M.L., 'Geometry of full wiped twin-screw equipment', *Polymer Engineering and Science*, September 1978 **18** 12 973–84.

24. MARTELLI F. G., *Twin-screw extruders: a basic understanding*, New York, Van Nostrand Reinhold Company, 1982.
25. TAYEB J., *Approche théorique et expérimentale de la cuisson-extrusion de l'amidon (PhD)*, Paris, Ecole des mines de Paris, 1986.
26. BARRÈS C., *Transformation de produits céréaliers par cuisson extrusion – Approche expérimentale et modélisation (PhD)*, Paris, Ecole des mines de Paris, 1990.
27. NOÉ I., *Approche expérimentale et théorique de l'extrusion bi-vis – Rapport de DEA*, Paris, Ecole des mines de Paris, 1988.
28. SZYDLOWSKI W., BRZOSKOWSKI R. and WHITE J. L., 'Modelling flow in an intermeshing co-rotating twin-screw extruder: flow in kneading discs', *International Polymer Processing*, 1987 **1** 4 207–14.
29. SZYDLOWSKI W. and WHITE J. L., 'An improved theory of metering in an intermeshing co-rotating twin-screw extruder', *Advances in Polymer Technology*, 1987 **7** 2 177–83.
30. WHITE J. L. and SZYDLOWSKI W., 'Composite models of modular intermeshing co-rotating and tangential counter-rotating twin screw extruders', *Advances in Polymer Technology*, 1987 **7** 4 419–26.
31. WHITE J. L. and CHEN Z., 'Simulation of non-isothermal flow in modular co-rotating twin screw extrusion', *Polymer Engineering and Science*, 1994 **34** 3 229–37.
32. CARDENAS-CAROTI R., GROSCLAUDE G. and LERAT L., 'Instrumentation d'un extrudeur bi-vis (31–38)', *Cuisson-Extrusion Colloque No. 41*, Nantes, INRA Editions, Paris, 1986.
33. MOHAMED I. O., *Modeling shear rate and heat transfer in a twin screw co-rotating food extruder (PhD)*, Michigan State University, 1989.
34. VAN ZUILICHEM D. J., VAN DER LAAN E., STOLP W. and VAN'T RIET K., *Food Extrusion Science and Technology: Modeling of heat transfer in a co-rotating twin-screw food extruder* (149–64), New York, Marcel Dekker, 1992.
35. VAN LEEUWEN J., GOREN M., SGRIMBERGEN A. A. and MOLJIN J. C., 'Enige aspecten van de beheersing van massa-temperaturen bij de verwerking van thermoplasten – DEEL I', *Plastica*, 1964 **6** 269–77.
36. VAN LEEUWEN J., GOREN M., SGRIMBERGEN A. A. and MOLJIN J. C., 'Enige aspecten van de beheersing van massa-temperaturen bij de verwerking van thermoplasten – DEEL II', *Plastica*, 1964 **7** 319–26.
37. VAN LEEUWEN J., GOREN M., SGRIMBERGEN A. A. and MOLJIN J. C., 'Enige aspecten van de beheersing van massa-temperaturen bij de verwerking van thermoplasten – DEEL III (slot)', *Plastica*, 1964 **8** 371–86.
38. SCHLÄFFER W., SCHIJF J. and JANESCHITZ-KRIEGL H., 'Measurements of radial temperature profiles in a single-screw extruder', *Plastics & Polymers*, June 1971 193–9.
39. MOTTAZ J., *Analyse thermiqueet mécanique en extrusion bi-vis corotative: étude expérimentale et modélisation numérique (PhD)*, Marseille, Université de Provence, 1997.

40. MOSSO K., JEUNIK J. and CHEFTEL J.-C., 'Température, pression et temps de séjour d'un mélange alimentaire dans cuiseur-extruseur bi-vis. Influence des paramètres opératoires', *Cahier Scientifique et Technique – Industries Alimentaires et Agricoles*, 1982 **99** 5–18.
41. FLETCHER S.I., RICHMOND P. and SMITH A.C., 'An experimental study of twin-screw extrusion-cooking of maize grits', *Journal of Food Engineering*, 1985 **4** 291–312.
42. DELLA VALLE G., TAYEB J. and MELCION J.P., 'Relations des paramètres opératoires avec la température et la pression de l'amidon pendant la cuisson-extrusion (39–49)', *Cuisson-Extrusion Colloque No. 41*, Nantes, INRA Editions, Paris, 1986.
43. DELLA VALLE G., TAYEB J. and MELCION J.P., 'Relationship of extrusion variables with pressure and temperature during twin screw extrusion cooking of starch', *Journal of Food Engineering*, 1987 **6** 423–44.
44. NOÉ I., *Etude des écoulements de polymères dans une extrudeuse bi-vis corotative (PhD)*, Paris, Ecole des mines de Paris, 1992.
45. TADMOR Z. and KLEIN, I., *Engineering principles of plasticating extrusion*, New York, Van Nostrand Reinhold, 1970.
46. JANSSEN L.P.B.M., *Twin screw extrusion*, Amsterdam, Elsevier, 1978.
47. TAYEB J., VERGNES B. and DELLA VALLE G., 'Theoretical computation of the isothermal flow through the reverse screw element of a twin-screw extrusion cooker', *Journal of Food Science*, 1988 **53** 2 616–25.
48. JEPSON C.H., 'Future extrusion studies', *Industrial and Engineering Chemistry*, 1953 **45** 5 992–3.
49. MOHAMED I.O. and OFOLI R.Y., 'Prediction of temperature profiles in twin-screw extruders', *Journal of Food Engineering*, 1990 **12** 2 145–64.
50. VAN ZUILICHEM D.J., VAN DER LAAN E. and KUIPER E., 'The development of a heat transfer model for twin-screw extruders', *Journal of Food Engineering*, 1990 **11** 187–207.
51. DELLA VALLE G., BARRÈS C., PLEWA J., TAYEB J. and VERGNES B., 'Computer simulation of starchy products' transformation by twin-screw extrusion', *Journal of Food Engineering*, 1993 **19** 1–31.
52. POTENTE H., ANSAHL J. and KLARHOLZ B., 'Design of tightly intermeshing co-rotating twin screw extruders', *International Polymer Processing*, 1994 **9** 1 11–24.
53. POTENTE H. and HANHART W., 'Design and processing optimization of extruder screws', *International Polymer Engineering and Science*, 1994 **34** 11 937–45.
54. VERGNES B., 'Analyse et modélisation du procédé d'extrusion BIVIS (61–67)', *40 ans d'extrusion bi-vis chez CLEXTRAL*, Firminy, Clextral, 1996.
55. TAYEB J., DELLA VALLE G., BARRÈS C. and VERGNES B., *Food Extrusion Science and Technology: Simulation of transport phenomena in twin-screw extruders* (41–70), New York, Marcel Dekker, 1992.
56. TAYEB J., DELLA VALLE G. and VERGNES B., *La cuisson extrusion: Modélisation thermomécanique du procédé d'extrusion bi-vis; Exemples*

d'applications (489–536), Paris, Lavoisier Tec&Doc, 1994.
57. VERGNES B., BARRÈS C. and TAYEB J., 'Computation of residence time and energy distributions in the reverse screw element of a twin-screw extrusion-cooker', *Journal of Food Engineering*, 1992 **16** 215–37.
58. DELAMARE L., *Evolutions morphologiques d'un mélange de polymères incompatibles dans une extrudeuse bivis corotative (PhD)*, Paris, Ecole des mines de Paris, 1995.
59. MOHAMED I.O. and OFOLI R.Y., 'Prediction of temperature profiles in twin-screw extruders', *Journal of Food Engineering*, 1990 **12** 2 145–64.
60. VAN ZUILICHEM D.J., ALBLAS B., REINDERS P.M. and STOLP W., *Thermal Processing and Quality of Foods: A comparative study of the operational characteristics of single and twin-screw extruders* (33–43), London, Elsevier, 1983.
61. VAN ZUILICHEM D.J., TEMPEL W.J., STOLP W. and VAN'T RIET K., 'Production of high-boiled sugar confectionery by extrusion-cooking of sucrose: liquid glucose mixtures', *Journal of Food Engineering*, 1985 **4** 37–51.
62. VAN ZUILICHEM D.J., JAGER T., DE RUIG J.A.J. and SPAANS E.-J., 'The influence of a barrel-valve on the degree of fill in a co-rotating twin-screw extruder', *Journal of Food Engineering*, 1989 **10** 241–54.
63. PETIT D., *Réduction de modèles et identification de modèles d'ordre réduit. Application aux processus de diffusion thermique (PhD)*, Marseille, Université de Provence, 1991.
64. PETIT D., SADAT H. and PEUBE J.L., 'Réduction de modèles de systèmes thermiques – Modélisation numérique en thermique', Ecole d'été GUT – CET – Institut d'Etudes Scientifiques de Cargèse, Institut Universitaire des Systèmes Thermiques Industriels, 1992.
65. HACHETTE R., *Réduction de modèle thermique par identification: Utilisation pour des conditions aux limites variables – Couplage à un modèle détaillé (PhD)*, Marseille, Université de Provence, 1995.
66. PETIT D., HACHETTE R. and VEYRET D., 'A modal identification method to reduce a high-order model: application to heat conduction modelling', *International Journal of Modelling and Simulation*, 1997 **17** 3 242–50.
67. PETIT D. and HACHETTE R., 'Model reduction in linear heat conduction: use of interface fluxes for the numerical coupling', *International Journal of Heat and Mass Transfer*, 1998 **41** 3177–89.
68. IDEL'CIK I.E., *Memento des pertes de charge – Coefficients de pertes de charge singulières et de pertes de charge par frottement. Collection de la Direction des Etudes et Recherches d'Electricité de France*, Paris, Editions EYROLLES, 1986.
69. CHEFTEL J.C., KITAGAWA M. and QUÉGUINER C., *La cuisson-extrusion: Nouveaux procédés de texturation protéique par cuisson-extrusion à teneur élevée en eau* (45–84), Paris, Lavoisier Tec&Doc, 1994.

5

Effective process control

C. J. Chessari and J. N. Sellahewa, Food Science Australia, North Ryde

5.1 Introduction

This chapter seeks to develop an increased awareness for the reader of the tools and strategies that may be used to maintain consistency of extruded product quality. Though by no means an exhaustive exploration of the subject, a range of issues will be covered to illustrate how control of product quality in an extruder may be achieved. The chapter covers:

- a basic review of control philosophy
- extruder control issues
- product quality issues
- instrumentation and sensors
- a case study in controlling extrusion product quality attributes.

A key issue to clarify right at the beginning is what we exactly mean by *quality*. This concept may mean different things to different people (and organisations). We shall define quality to mean the consistent reproduction of an item with predetermined specifications with minimal off-specification product. Therefore, the need we seek to satisfy is to maintain and/or move to different quality regions with minimal disruption.

The key features that the reader should take away from a reading of this chapter are the following:

- a working definition of process control
- the crucial relationship between sensors/instrumentation and control
- the strategy by which a *product quality* control system may be implemented.

5.1.1 What is process control?

A fundamental question that needs to be agreed upon is 'what is process control?' In this exercise, it is not appropriate (and not helpful) to go through the details given by automation suppliers and the detailed mathematical analyses of academics. At the basic philosophical level control is all about: *Keeping what you want where you want.* This statement makes no assertion about automatic control, manual control, feedback and feedforward systems, nonlinearities, or robustness, etc. However, these topics are all crucial in implementing a reliable system. It is important to recognise that you can control a process without any automation. For example, the rejection of product on lines can be done by visual inspection. The second key feature of this definition is that it makes no claim of *where* the process should be. This task is in the realm of optimisation. The best place to position your process will be determined by a compromise between everything affecting your process that satisfies your business objectives. In summary, control is a servant of optimisation and optimisation is a servant of the business objectives.

The fundamental features of a feedback control loop are illustrated in Fig. 5.1. In order to achieve high performance from the control system, all the components described below must be sufficient to satisfy the requirements of the system. For each component of the control system there are some key issues:

Process

- Require detailed knowledge of the process.
- Need to understand the dynamics of the process.

Measurement (sensors)

- A control system cannot work without good field measurements.
- May not be available.
- May be too slow.
- Soft sensors (inference of a desired variable using measured-related values).

Action

- Choose a design (robust, maintainable, flexible) that is the simplest to do the job.

Change agent

- Need to understand the fundamental limitations of the final control elements (e.g. valves, variable speed drives, hydraulic cylinders in robot arms).

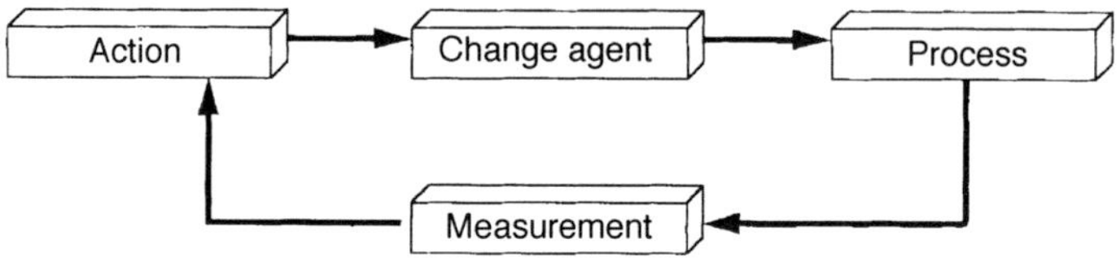

Fig. 5.1 General feedback control loop.

5.1.2 Why control?

The fundamental aim of any process is to convert the input feedstock into *desired* products using available resources in the most economical way. The process must satisfy the technical, economic and social requirements imposed upon it in the face of ever-changing external influences in the course of its operation. Among the requirements that need to be met are:

- *Safety.* Safety of personnel and equipment must be paramount.
- *Quality.* Customer requirements concerning quality and consistency must be met.
- *Environmental.* Must comply with prevailing environmental standards.
- *Economic.* Operation must conform to market conditions – the availability of raw materials and demand for final product. It should be as efficient as possible in the use of raw materials, energy, capital and personnel. Consequently, it must operate at conditions of minimum cost and maximum profit.

The aforementioned requirements necessitate continuous monitoring of the process and external intervention to guarantee that operational objectives (the business objectives) are met. Due to the increasing frequency of change in the market place there is a greater emphasis on the deployment of resources to extract the maximum profit from existing capital investment.

The implementation of appropriate process control technologies can be expected to deliver the following process benefits:

- improved quality
- consistent quality
- increased throughput
- reduction of energy usage
- reduction in waste
- more precise control
- more stable operation
- less operator intervention
- faster start-ups
- decrease in processing downtimes
- reduced product giveaway
- increased yield.

A number of studies have been undertaken to quantify the dollar benefits of such implementations. A Warren Centre study[1] estimated benefits equivalent to two to six per cent of operating costs. The range of industrial case studies involved in the evaluation were petroleum refining, petrochemicals, minerals, power generation, sugar refining, and sewage treatment. A benchmark study by ICI[2] indicated that the effective use of improved process control technology could add more than one third to the worldwide ICI Group's profits.

Very little in the way of benefits studies applicable to the food industry are available in the open literature. A study by White[3] based on several baking plants in the US before and after automation suggests that $200 000 per line per year can be realised. This study only considered economic benefits arising from reductions in machine downtime. A recent survey (1997) conducted in the US[4] has stated that the three top trends having most impact on manufacturing in the food industry in the next five years are:

1. automation/information management
2. flexible/more efficient manufacturing
3. supply chain management.

A sample of the drivers for change and benefits cited in this survey are presented:

- more precise and accurate technology at an acceptable cost
- real-time information from the production floor, which affects inventory, production planning and maintenance
- on-line analysis because immediate, relevant results will run the production process efficiently
- control system integration for greater production efficiencies
- a means of achieving more consistent product quality while reducing workforce loads and increasing flexibility.

5.1.3 Challenges in extrusion process control

As the awareness of food quality and the concern about food safety has increased, it has forced the food and feed industries to place more emphasis on the consistency of product quality. Food process control has therefore become of considerable importance in recent years.

In extrusion cooking, biopolymers, mainly starches and proteins, are plasticised with water and subjected to mechanical and thermal energy treatment to achieve the desired texturisation for food type end products and specific functional properties for modified starches and/or proteins. Figure 5.2 shows the interactions of raw material properties, process variables and product characteristics.

Control of extrusion processes is difficult due to strong interactions between mass, energy and momentum transfer, coupled with complex physico-chemical transformations, which govern final product properties. The primary extrusion process parameters include feed formulation, feed moisture content, feed rate, screw speed, barrel temperature and screw and die configuration. For most extruder applications, typically die pressure, die temperature, and motor torque (or current) are used as measured process outputs to monitor product quality indirectly.

A number of control strategies have been developed for the extrusion process.[5–8] An extrusion control system aims at controlling the product characteristics ($p_1 \ldots p_n$), as seen in Fig. 5.3 (reproduced from Kulshreshtha *et al.*[5]). However, *these cannot be measured on-line* and are known only after a delayed laboratory evaluation. One of the major difficulties with developing

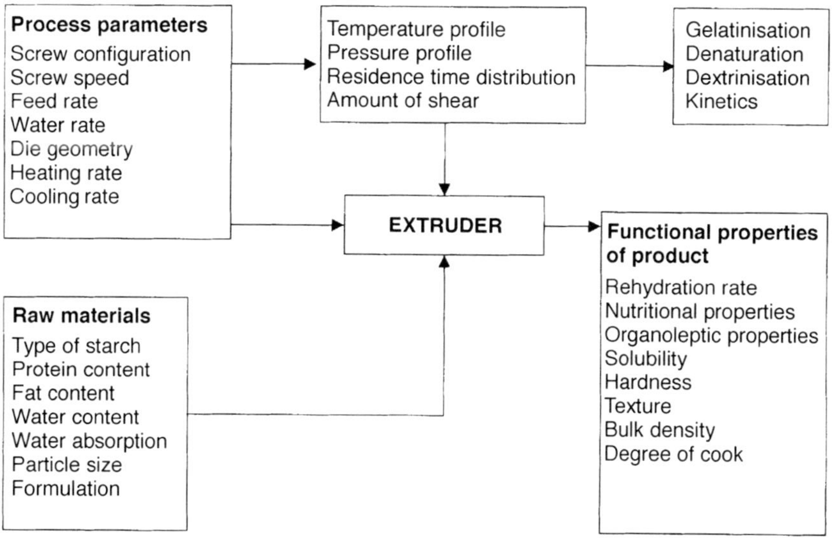

Fig. 5.2 Interaction of raw material properties, process variables and product characteristics.

advanced control strategies for cooking extruders is the lack of on-line product quality estimators. By using a correlation matrix (C), the product characteristics correlate to the product temperature (T_d), pressure (P_d) at the die, and the power needed to run the screw (E). These three variables are the process outputs, which can be measured on-line and are commonly used to monitor the extrusion process. Feed rate (FR), water addition rate (WR), screw speed (SS), and

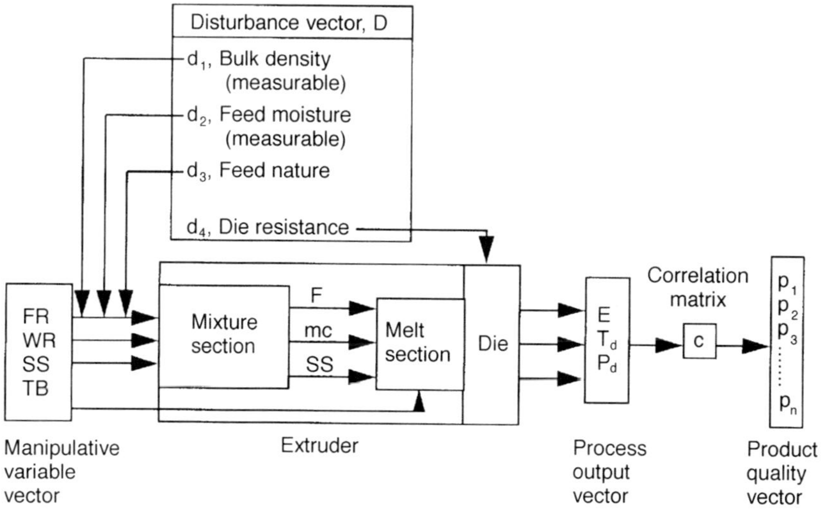

Fig. 5.3 Overview of food extrusion process. (Reproduced from Kulshreshtha, Zaro and Jukes (1991).)

temperature of the extruder barrel (TB), are the process input variables. The total feed rate through the extruder (F) is the sum of FR and WR. The resulting moisture content (mc) is a function of these variables. The disturbances ($d_1 \ldots d_n$) associated with the extrusion process relate mostly to changes in feed characteristics such as bulk-density, moisture content and composition.

5.2 Product requirements

There are two main product requirements. These are:

- consumer requirements
- regulatory requirements (weight of product in the package, composition, nutritional claims).

The consumer requirements consist of sensory attributes such as taste, colour, texture, size, shape, 'bowl life' (for breakfast cereals), bulk density and microbiological safety.

Most of the consumer and regulatory requirements are affected by the composition of the raw materials, extruder operating conditions and post extruder operations such as drying, toasting and flavour addition. The key control points in meeting product quality will be discussed in section 5.3.

5.2.1 Flavours and colours

Compared to cooking by conventional thermal processes such as jacketed batch cookers or continuous rotary steam cookers, the residence time in an extruder is much less (seconds compared to minutes). Therefore, extrusion cooking is essentially a short time high temperature process and a number of traditional 'cooked' flavours are not produced. As in traditional cooking processes, flavour is generated during extrusion cooking by a number of reactions that take place which are controlled by the composition, temperature and residence time. However, in extrusion cooking these reactions are accelerated because of shear forces. A number of the flavours and colours of extruded products are formed by the reaction of amino acids and reducing sugars (Maillard reaction). Recent work[9, 10] suggests that adding flavour precursors could control these flavours. The kinetics of these reactions are complex and not fully understood.

Generally, products coming out of extruders do not have strong flavours because of the short residence time and the flashing off of the volatile flavour compounds as the product expands. Therefore, flavours are usually added after extrusion to make the product more palatable. Most extruded snack products are dried and flavours are added by spraying an oil emulsion incorporating flavours. Breakfast cereals are generally toasted after extrusion by subjecting the product to high temperatures. As most breakfast cereals contain sugar, some caramelisation takes place, which introduces flavours as well as introducing a dark colour. Flavours and colours are also developed during toasting from the

Maillard reaction. Sometimes a sugar solution is sprayed to breakfast cereals before toasting to develop stronger flavours. Another method of enhancing flavours is by adding ingredients such as malt flour to the formulation.

5.2.2 Texture

Texture is one of the most important sensory attributes of extruded products. The rapid flashing of water forms the characteristic texture of extruded products when the starch melt comes out of the extruder die. As the pressure is suddenly reduced from a high pressure in the extruder to atmospheric pressure, water changes from liquid to vapour. As bubbles of water vapour come out of the starch melt, the product stretches, and the matrix sets because of evaporative cooling. Air bubbles get trapped in this matrix and the characteristic puffed structure is formed. The texture of extruded products is dependent on the cell size distribution and cell wall thickness. Several studies have been carried out to understand the relationship between the structure and texture of extruded products.[11,12] Studies have also been carried out in modelling the expansion of extruded products.[13]

The rheology of the melt has a significant effect on the expansion mechanism and therefore on the final product texture. The rheological properties of the melt are controlled by the formulation, temperature profile, barrel moisture content, screw speed and screw profile (which affects shear forces in the extruder) and the die design. The interaction of carbohydrates, proteins and lipids has an effect on the melt rheology and product texture. The effect of ingredients on the physical/structural properties of extrudates is reported by Moore *et al.*[14] The product texture is also affected by the die design and post extrusion operations (such as drying and toasting).

5.2.3 Bulk density

The bulk density is a very important product quality attribute from the viewpoint of commercial production of extruded products because most extruded products are filled by weight and not by volume. Therefore, if the bulk density varies during production, either the pack will not be full or it will overflow. As both these scenarios have serious production implications, in addition to the moisture content, bulk density is a quality attribute that is measured regularly for quality assurance purposes. In addition to controlling the correct volume of product in the pack, if the bulk density is controlled properly, generally the product texture will be within the required quality limits. This is because there is a relationship between the bulk density and the texture because both these parameters are controlled by the degree of expansion. The die design also has an effect on the bulk density.

5.2.4 Size and shape

Today there are a variety of extruded products of different shapes and sizes. Shapes and sizes are specified to fulfill marketing requirements and very often the shape and size are the only changes of many of the new extruded snack products that are launched. The shape and size are determined mainly by the design of the die and the rotating knife and the speed of rotation of the knife. However, the shape and size are also affected by the melt rheology, which in turn is controlled by the formulation and operating conditions.

5.2.5 Microbiological quality

One of the most important consumer requirements is the microbiological safety of the product. Most conventional extruded products such as snack foods and breakfast cereals are safe to eat because the raw materials are subjected to high temperatures (> 130°C) and the water activity of the product is low because the product is dried to a moisture content of less than five per cent. However, the microbial safety can be an issue when products for human and animal consumption are made with raw materials such as offal and other animal waste products, which can have pathogens and a high microbial load. The microbiological safety can also be a cause for concern with some foods that have a high water activity. Textured products made with animal or plant proteins containing high moisture are examples.

Although it is well known that most vegetative organisms, yeast and moulds are destroyed under typical extrusion conditions, the operating conditions under which spores are inactivated are not well understood. Queguiner *et al.*[15] were able to achieve a four to five log decrease of *Streptococcus thermophilus* using low shear conditions and barrel temperatures above 130°C. Bulut *et al.*[16] studied the effect of high shear forces at low temperature (75°C) at a moisture content of 19% and obtained a five log reduction of *Microbacterium lacticum*. In a recent feasibility study,[17] it was found that the inactivation of *Bacillus cereus* is caused mainly by thermal effects. However, shear forces appear to enhance this effect. Therefore, providing a temperature of at least 130°C is reached in the extruder, the safety of the product is assured. Like in any other food processing operations, it is important to have other procedures such as good manufacturing practices and HACCP (Hazard Analysis Critical Control Points) plans to ensure that the product is not contaminated during post extrusion operations.

5.3 Key control points in meeting product requirements

The quality of extruded products is controlled by the formulation, pre-extruder operations (blending and preconditioning), extruder screw configuration, die design, extruder operating conditions (feed rate, screw speed and the water injection rate into the barrel) and post extruder operations (drying, toasting and flavour addition). An overview of the key control points in meeting product requirements is shown in Fig. 5.4. Although these specific control points affect product quality, it is important to treat the extrusion process as an integrated system and ensure that all control points are properly controlled.

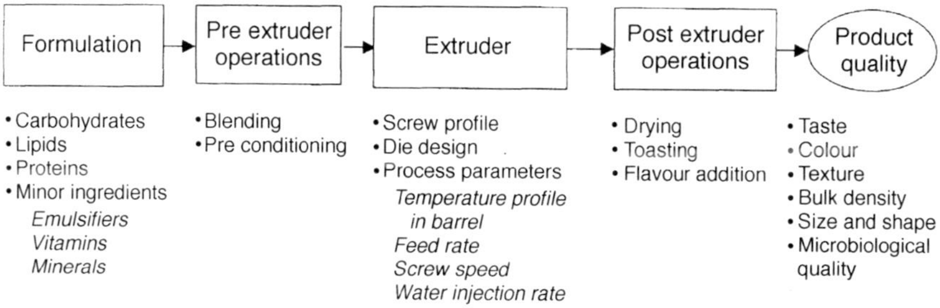

Fig. 5.4 Key control points in meeting product requirements.

5.3.1 Manipulated variables

In an extrusion system, for a given formulation, there are four main categories of variables, which affect product quality. These are:

- the extruder screw configuration and length to diameter ratio
- die design
  - diameter
  - shape
  - land length
  - number of dies on die plate
- design of rotating knife
- process variables that the operator can change during the run.

The extruder screw configuration affects the degree of mixing, shear forces introduced, amount of heat generated by friction and the residence time distribution (RTD). All these factors affect the degree of cooking, the melt rheology and product quality and uniformity. The length to diameter ratio affects the residence time and therefore the degree of cooking.

Although most of the work in an extruder takes place in the extruder barrel where the raw materials are transformed, the die design has a significant effect not only on the product size and shape, but also on the die pressure, melt rheology (as the melt flows through the die), the pressure drop across the die, barrel fill and the amount of energy generated in the extruder. All of these factors affect the product quality and consistency. The design of the rotating knife and its speed affects the size and shape of the product.

Although the extruder screw configuration, length to diameter ratio, die design and the rotating knife design all affect the product quality, none of these variables could be changed when the extruder is running. However, the following process variables could be changed when the extruder is operating:

- screw speed
- water injection rate
- feed rate
- temperature profile in the barrel
- speed of rotating knife.

The screw speed affects the residence time, the amount of shear introduced, the melt viscosity, amount of frictional energy generated and the barrel fill. The water injection rate determines the barrel moisture content, which in turn controls the frictional energy generated, and the melt viscosity. The feed rate controls the energy generated, the residence time and the barrel fill. The temperature profile of the barrel affects the product temperature, which in turn affects degree of cook and the melt viscosity. The barrel temperature is controlled by external heating systems (electrical conduction heating, hot oil, steam or electrical induction, depending on the type of extruder). In some extruders, thermal energy is added to the product by direct steam injection. If necessary, heat could be removed from the extruder barrel by circulating cooling water.

5.3.2 Controlled variables

There are four main controlled variables in an extruder. These are:

- specific mechanical energy
- die melt temperature
- die pressure
- flow rate through the die.

The specific mechanical energy (SME) is defined as follows.

$$\text{SME} = \frac{\text{Total energy}}{\text{Flow rate}}$$

If there is no external heat applied, the total energy is equal to the heat generated by friction. Otherwise, the external heat applied should also be included.

As mentioned in section 5.3.1, the die melt temperature (and the die pressure) can be controlled from the manipulated process variables. The flow rate through the die can be calculated by dividing the total flow rate through the extruder by the total cross-sectional area of the die.

5.3.3 The importance of good design and controlling the key control points

The objective of controlling an extrusion system is to control the product quality and the consistency of quality. Before attempting to control the extruder, it is important to design the whole extrusion system (including pre- and post-processing) correctly. The extrusion process must be regarded as an integrated process involving formulation of raw materials, pre-processing, processing in the extruder (including the die) and post extrusion processing. In order to obtain the required product quality, the whole system has to be properly controlled.

As far as the extrusion process is concerned, it is generally true that for a given formulation, screw configuration and die design, if the main process variables (feed rate, screw speed, water injection rate, barrel temperature profile and the speed of the rotating knife) are controlled properly, the controlled variables (SME, die temperature, die pressure and flow rate through the die) will be maintained at the desired values. This will produce a good product consistently.

One of the most common causes of non-uniform product is instability in the extruder. This is caused by the emptying of the barrel. This happens when the screw speed is too high relative to the feed rate. Therefore it is important to control the feed rate and screw speed properly so that the barrel is not empty or flooded (a phenomenon that occurs when the screw speed is too low relative to the feed rate). It is recommended that an accurate gravimetric powder feeder be used so that a constant mass flow rate is obtained even if the bulk density of the raw materials varies. It is equally important to have a good pump to inject water into the extruder barrel because even a small fluctuation in the water rate could have a significant impact on product quality.

5.4 Instrumentation

Measurement of key variables is essential for control. Without these measurements there can be no control. Typically, extruder instrumentation consists of an ampere meter to estimate drive motor power, simple solid and liquid ingredients feeding systems and associated flow meters, a thermocouple to measure die melt temperature and observation at the discharge end to estimate product quality.

The amount of instrumentation that is appropriate for any particular extrusion application depends on the economics of the product and the ability to use the information that is generated from the instruments. More sophisticated extrusion systems will have the feed and liquid ingredients tightly monitored and controlled. In addition, a range of thermocouples along the barrel and at the die assembly may be encountered. In some cases, a pressure transducer will be found at the die. A very brief review of existing sensors is given in section 5.4.1. Any number of instrumentation suppliers will be able to provide reams of information about these instruments. Our focus will be on novel sensors, which offer something different from the standard instruments.

5.4.1 Existing sensors

Powder feed rate

Powder feed rate is typically measured by a volumetric feeder. That is, the volume of material fed to the extruder is maintained at a specific amount. However, slight feed rate disturbances due to bulk density changes can cause significant disruption to the extruder. There are a number of alternative volumetric feeder types:

- *Single screw feeder.* Most common, volumetric rate is proportional to screw speed.

- *Disk feeder.* Material delivered is a function of rotational speed. Flooding of free flowing material is minimal.
- *Volumetric belt feeder.* Feed rate is controlled with the speed of the belt.

In practice, a gravimetric feeder – where a specified mass of material per unit time is fed to the extruder is a safer alternative. These units are more expensive but will lead to more stable extruder operation. The two common types of gravimetric feeder are:

- loss in weight feeders
- weigh belt feeders.

Both measure the mass rate and adjust automatically for fluctuations in bulk density and flow properties.

Liquid feed rate
Liquid feed rate can be measured by a number of instruments:

- *Rotameter.* Simple, not amenable to automatic control.
- *Orifice and venturi meters.* Measures pressure drop, which is proportional to flow, suitable for water and steam.
- *Positive displacement meter.* Flow is measured by the displaced volume, suited for clean materials and where high accuracy required, for example in addition of enzymes.
- *Magnetic flow meter.* Presents no obstruction to flow, suitable for conductive liquids, slurries and viscous solutions.
- *Metering pumps.* For small quantities, volume of fluid depends on speed of rotation or length of pump stroke.

Screw speed
Screw speed is typically measured by:

- *Direct current tachometer.* Voltage output is directly proportional to speed of rotation.
- *Pulse generator.* Transducer producing an external magnetic field that generates an alternating current. The screw speed is proportional to the frequency of the current.

Torque
The most common transducer uses strain gauges to measure angular deflection, which is directly related to torque.

Barrel, die and melt temperatures
Temperatures of the barrel and die block are measured by inserting thermocouples (J or K type) along the length of the extruder. The measurement of the melt in the barrel is difficult because it is not possible to protrude thermocouples into the melt. The melt temperature in the die can be measured by inserting a thermocouple into the melt.

Product temperature (pyrometry)
It is possible to measure the temperature of the product exiting the die by using infrared non-contact temperature measurement systems.

Die pressure
Pressure transducers placed in the die plate can be used to measure the pressure at the die. This measurement can be used to monitor the extruder. The common transducer is a sealed force transducer. In practice, pressure transducers tend to suffer from abrasion and need attention with respect to maintenance, calibration and temperature compensation.

Melt rheology
The rheology of an extrusion melt is an important material characteristic. There is currently no well-accepted method of on-line measurement of the rheology of an extrusion melt. There are a number of off-line tools available to measure rheological behaviours. One can choose from a capillary viscometer, Carri-Med rheometer, Rapid Visco Analyser (RVA) or the recently developed Micro Fourier Rheometer. All these instruments impose their own standard way of moving the sample and measuring the resultant resistant force.

5.4.2 Novel sensors

In discussing novel sensors, the focus is upon measuring product properties, which are deemed important to control (quality measures such as bulk density and texture) and are difficult to measure on-line.

Acoustics
A promising development area is the recent emergence of acoustics-based estimation techniques.[18] The basis of this approach relies on the fact that as the extrudate leaves the die, the water content partially flashes, causing the extrudate to expand. There are discernible differences in sound as this steam escapes when different products are being extruded. It is the interaction between this expansion, the rheology of the extrudate, and the resultant bubble size distribution that gives the final product many of the qualities that both the manufacturer and the consumer seek. As the product exits the extruder, a characteristic 'popping' sound is clearly evident. A microphone connected to a personal computer (PC) with a sound card is used to record digitally the sounds emitted by the extruder. This data can be used to predict various quality attributes. A typical configuration for capturing the acoustic emissions from an extruder is shown in Fig. 5.5.

Near infra red (NIR) spectroscopy
NIR spectroscopy has been used extensively in the agricultural and food industries for the past thirty years for non-invasive, hygienic and safe routine measurement of protein, fat and moisture. This involves the measurement of the

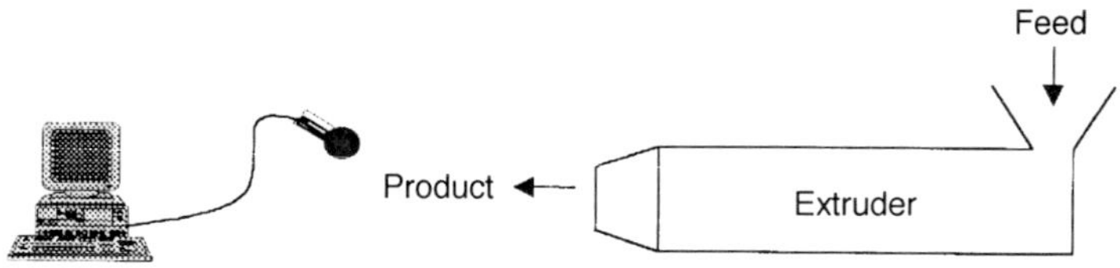

Fig. 5.5 Configuration for capturing acoustic emissions from an extruder.

wavelength and intensity of the absorption in the electromagnetic spectrum from 400 to 2500 nm. This is correlated to an analytical or quality aspect of a sample. NIR is energetic enough to excite overtones and combinations of molecular vibrations to higher energy levels. NIR spectroscopy is typically used for the measurement of organic functional groups, C—H, O—H, N—H, and C = O. It is especially applicable to the study of the chemical changes undergone by starches during the extrusion process.

A collaborative project between Food Science Australia, CSIRO Plant Industry, BRI Australia Ltd, and FOSS NIR Systems recently has extended the work carried out in Europe in the use of NIR spectroscopy for characterising material transformations within cooking extruders.[19,20] In this study, two probes were inserted into the extruder melt and the absorbency of NIR spectra was correlated with physical, textural and sensory properties of the product.

Electronic nose

Electronic noses are sophisticated sensors that create digital fingerprints of smells. Measurement by the electronic nose is objective, repeatable, highly accurate and relatively cheap. Interpretation is simple, quick, and in real-time. Like the human sense of smell, the electronic nose learns by experience and improves the more it is used. It is designed to analyse, recognise and identify volatile chemicals at low (parts per billion) levels. The technology is based on the absorption and desorption (passing through) of volatile chemicals onto an array of sensors, which exhibit specific changes in electrical resistance, measurable across each sensor element, on exposure to different odours and aromas.

Electronic tongue

The human tongue can distinguish between a broad array of subtle flavours using a combination of just four elements of taste (sweet, sour, salt and bitter). Researchers at the University of Texas in Austin have designed an electronic tongue using chemical sensors.[21] A possible use in extrusion technology is to use the electronic noses and tongue to obtain reliable instrumental sensory evaluation without the expense of training and maintaining sensory evaluation panels.

Soft sensors

A soft sensor is essentially a strategy for developing sensors that can be inferred from a combination of existing sensor information. Essentially, it facilitates development of non-contact, non-invasive sensors of product quality, which are exactly the type of sensors that are advantageous for use in the food and feed industry. Let us clarify by an example. It is not possible to measure directly the specific mechanical energy (SME) of an extrusion process. However, it is possible to measure the screw speed, motor torque and feed rate of an extruder. These process variables can be measured easily on-line. Knowing the relationship between SME and these variables allows us to develop a computed (virtual) sensor for SME. In section 5.6, we illustrate a more complex soft sensor where bulk density and moisture content of the extrudate can be measured on-line from available process measurements.

5.5 Process monitoring

As discussed in section 5.4, it is necessary to have appropriate instruments to measure product quality attributes and process variables. Signals from these instruments could be used as inputs to a process control system. However, before control systems are developed, it is possible to use the information from the instruments to monitor the process. In process monitoring, real time information is displayed on a visual display unit (VDU) and the data is logged for analysis. The process operator uses real time information to make decisions and to control the process manually to minimise the generation of out of specification products. The logged data can usefully be used to analyse the performance of the plant.

A variety of affordable software for data logging is readily available. The most common software that is used for process monitoring also has some supervisory control capability. These are called SCADA (Supervisory, Control and Data Acquisition) systems. These software systems are now widely used in the food industry because they are easy to configure, are user-friendly, reliable, run on personal computers (PCs), are easy to interface with most instruments and control systems and are relatively cheap. Many new extruders include a SCADA system. These can be used for real time process monitoring and can be linked to a more advanced control system. Many older extruders could be connected to a SCADA system through a suitable interface.

5.5.1 System architecture

When designing a SCADA system, the following should be taken into account:

- Ease of interfacing to the instruments and local control system.
- The system should provide supervisory control as well as process monitoring capability.
- 'Open' architecture, so that the system could be changed easily. It should be possible to change the SCADA software without carrying out major changes

to the hardware and electrical wiring. The system should be independent of software vendors.

- It should be possible to operate the extruder using a 'local' control panel so that if the PCs running the SCADA and other control software were to fail, the extruder could still be run. It is therefore recommended to have a 'local'/'remote' switch so that the extruder could be controlled locally from a control panel or remotely from a supervisory computer.
- All the safety interlocks should be either hard wired or controlled through a Programmable Logic Controller (PLC) and not through the SCADA computer. This is because PLCs are more reliable in an industrial environment than a PC.
- The system should be expandable (adding extra sensors, local controllers).

A typical system to satisfy the above conditions would consist of local controllers (for example dedicated temperature controllers, motor controllers), a PLC, a PC for running SCADA software (essentially for process monitoring) and another 'control' PC that is networked to the 'monitoring' PC. New set points would be calculated (for instance using a predictive model of the extruder) in the control computer when there is a disturbance in the extruder. The new set points are downloaded to the local controllers (e.g. water injection pump motor controller) via the process monitoring computer. In this architecture, the PLC acts as an interface between the extruder and the process monitoring PC. All the control is carried out in the local controllers. An overview of such a system is shown in Fig. 5.6.

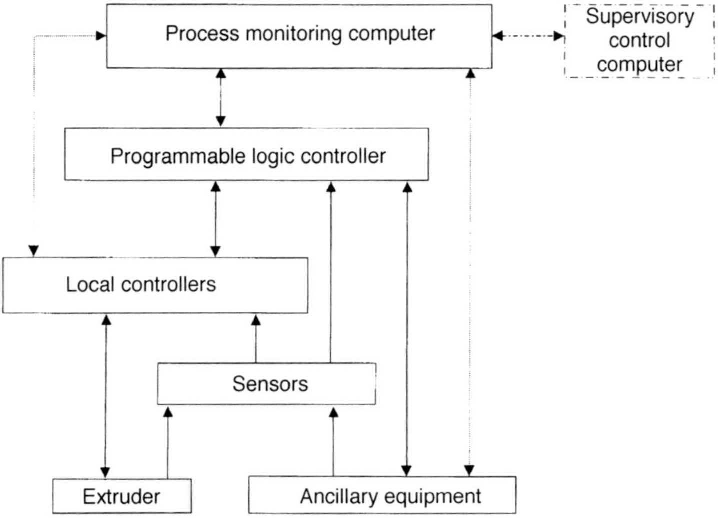

Fig. 5.6 Overview of a typical process monitoring and supervisory process control system.

5.5.2 Features of a process monitoring system

A good process monitoring system should have the following features:

- ease of configuration and use
- real time and historical trends
- alarms
- report generation
- ability to calculate key parameters from measured variables (e.g. SME from motor torque and feed rate)
- downloading data to spreadsheets and databases
- statistical process control (SPC) capability
- SCADA software must easily interface with well known PLCs and local controllers.

One of the most useful features of a process monitoring system is the ability to view real time and historical trends. This feature enables the operator to carry out 'cause and effect' studies and to make decisions. For example, the operator may notice an increase in the die melt temperature and a corresponding decrease in the water injection rate (caused by a faulty pump, for example). Therefore, the operator could identify the problem and take prompt corrective action.

Modern SCADA packages have good alarm facilities and graphics. If critical process variables reach values outside the normal range, the system could be configured to generate alarms to draw the attention of the operator. These alarms could be audible or changes of colour on the VDU. Most SCADA packages have SPC capability where critical process parameters could be trended in real time and continuously compared with upper and lower control limits. Printouts of such trends will generate the required documentation for quality assurance purposes within a good manufacturing practice system.

The ability to download data to spreadsheets and databases is another useful feature of a SCADA system. The analysis of vast amounts of data increases the understanding of the system by the operator and other technical personnel. Process understanding is an essential precursor to the design of a good process control system. The generation of reports (time or event driven or on demand) is useful for production managers. These reports can give alarm logs and other useful information such as amount of product made, amount of raw materials used, yields, deviation from targets, etc.

5.5.3 Designing and implementing a process monitoring system

The success of a process monitoring system depends on its proper design and implementation. In the design of the system it is very important to involve the end users (operators, supervisors and technical personnel). The designer must ensure that the requirements of the end users are considered right at the outset. Items such as the man machine interface (including screen layouts, colours, and graphics), alarm requirements, reports, groupings of trends have to be discussed with the end users and then a user specification should be written. Essentially, the user specification is a document that gives details of all the requirements of the end user. It is then necessary to write a functional specification. This specification describes exactly how the system will be designed after taking into account the constraints of the software, time scales and budgets. The user should be consulted during the preparation of the functional specification and the user should agree to the functional specification before the system is configured. The user should also be consulted during the configuration stage to get feedback on the screen layouts, colours, etc, so that minor modifications could be carried out as necessary. The system is then tested and the users are trained before final handover of the system. Good documentation (operating instructions and troubleshooting guide) is also important for the successful implementation of the system.

A well-designed process monitoring system is an extremely useful tool enabling an extruder operator to run the extruder under optimum conditions. Such a system will result in improved product quality and reduced waste without using a sophisticated control system.

5.6 Process control in action

The fundamental control problem as we stated in section 5.1 is that we want to keep what we want where we want.

5.6.1 Implementation procedure

There is a well-established process for implementing process control and optimisation strategies:

1. establish a team
2. establish a base case
3. prepare an opportunities list
4. quantify the benefits
5. prioritise the opportunities
6. select and design a suitable solution
7. implement the solution
8. audit the implemented solution.

Successful project outcomes have resulted from adhering to this standardised approach.

Laws of control and optimisation

These two simple rules are the result of many years of experience from many practitioners in the field.

1. The simplest system that will do the job is the best. (Principle of parsimony.)

2. The process must be understood. (Principle that imposed complexity cannot circumvent the fundamentals of the process).

5.6.2 Review of extruder control

The aim of extruder control is usually to compensate for steady state disturbances and maintain product quality. However, start-up, shutdown and changeover procedures may also result in a large amount of waste. The main steady state disturbances are associated with fluctuations associated with the viscoelastic nature of the material, barrel pressure distribution irregularities, water pressure variations (which causes variations in the water injection rate), feed rate variations or variations in raw material composition and moisture content. A number of control strategies have been developed for the extrusion process.[22–28] Table 5.1 presents a review of recent work in the area of food extrusion control.

Despite a large amount of work in the area of controlling extruders, there is no standard approach that has been applied successfully in industry. The main reason for this is primarily because of the complex nature of the extrusion process, which does not lend itself easily to traditional control approaches. The current state of art is to control tightly the dry and liquid feed ingredients and maintain consistent specification on raw material composition. However, as previously stated, even at steady state operations, there are a range of disturbances that affect the process and consequently product quality.

5.6.3 Development of an extruder product quality controller

We will illustrate this topic by a case study. One class of control strategy that can be used for control problems with interactions or complex responses such as found in the extrusion process is model predictive control (MPC).[30] The basis of model predictive control is a dynamic model that describes the features of the process. The model is used to predict the future trajectory of the process outputs for any pattern of input variation. Models are usually linear and obtained by direct identification from plant data. The controller is developed by using the model and an optimisation procedure to select the best way to manipulate the process inputs in order to achieve the control objectives. The *Connoisseur*™ [31] model predictive control software package was used to develop and implement the controller.

In this example, we used a soft sensor approach to control the moisture content and bulk density of a simple expanded extruded product simultaneously.[29, 32] The transformation of maize flour to an expanded product was investigated using Food Science Australia's fully instrumented APV-Baker MPF40 co-rotating twin-screw extruder.

Dynamic analyses have shown that for maize flour, SME and motor torque correlate well to the dynamic responses of bulk density and moisture content. The control approach used was to select a desired bulk density and moisture

Table 5.1 Review of recent work in control of extruders

Source	Control strategy	Manipulated variable	Controlled variable	Extruder	Approach	Control result
Moreira *et al.* (1990)[25]	Adaptive feedback/ feed forward	Screw speed	Die pressure	Baker-Perkins (50 mm) twin-screw	Simulated/ Experiment	Simulated
Hofer and Tan (1993)[6]	Feedback/feedback control system with disturbance prediction	Heating input, cooling rate and SME	Extrudate temperature	APV-Baker MPF 50/25 twin-screw	Experimental	Simulated
Lu *et al.* (1993)[24]	Model-based control	Screw speed, moisture content, feed-rate and barrel temperature	Die pressure and motor torque	APV-Baker (50 mm) twin-screw	Experimental	
Singh and Mulvaney (1994)[7]	Two single loop PID controllers without decouplers	Screw speed and barrel heating	Product temperature and motor torque	APV-Baker MPF twin-screw	Experimental	Simulated
Eerikainen *et al.* (1994)[28]	Feed forward artificial neural network	Liquid flow, screw speed and feed-rate	SME, pressure and torque	Continua 58, Werner & Pfleiderer twin-screw	Experimental	Simulated
Schonauer and Moreira (1995)[26]	GPC	Feed-rate and screw speed	Product colour and moisture content	APV-Baker MPF twin-screw	Experimental	Simulated
Kulshreshtha *et al.* (1995)[23]	PI feedback/model based set point adjustment	Feed-rate and screw speed	Die pressure		Simulated	Simulated
Haley (1998)[22]	MPC	Screw speed	Bulk density (using SME)	Wenger (25 mm) twin-screw	Experimental	Simulated
Altner (1999)[29]	MPC	Feed-rate, screw speed and barrel moisture	Bulk density (using die pressure)	APV MPF 40 twin-screw	Experimental	Experimental

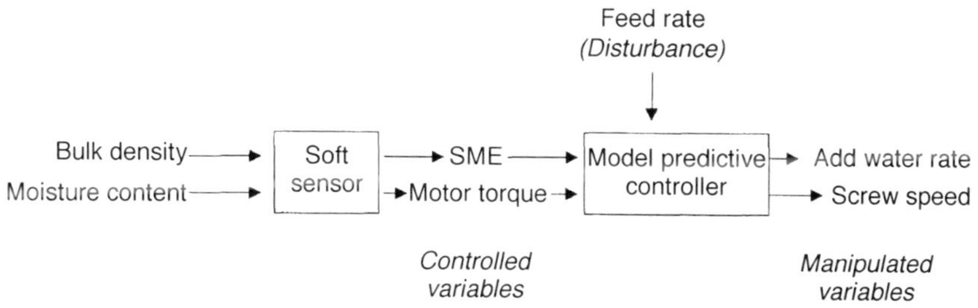

Fig. 5.7 Overview of controlling product bulk density and moisture content by linking a soft sensor with a model predictive controller.

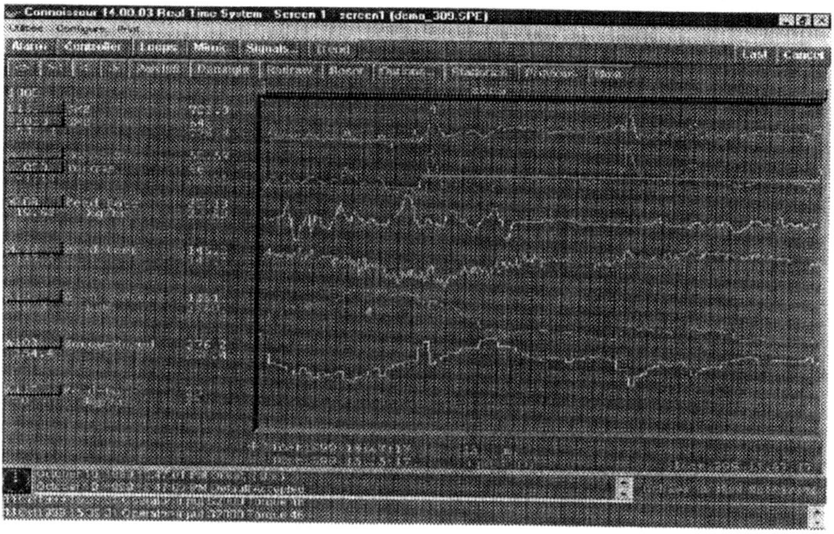

Fig. 5.8 Screen capture of a model predictive control system to control bulk density and moisture content of the extrudate.

content and estimate the required SME and motor torque using a soft sensor. The model predictive controller then adjusts the screw speed and added water rate to give the required SME and motor torque (Fig. 5.7). In this example the feed rate is a disturbance.

The experimental results showed that it is possible to specify a desired bulk density and moisture content, and for the control system to achieve the required values for these product quality attributes. A screen capture of the system (Fig. 5.8) shows that the system settles to the desired bulk density and moisture content within three minutes.

These results show that model predictive control in conjunction with a soft sensor approach to obtaining product quality relationships is a tool which has promise for controlling quality attributes of extruded products. Further work is focused on developing soft sensors for textural and sensory quality attributes.

5.7 Summary

In summary, process control seeks to assist the fundamental aim of any process – to convert the input feedstock into desired products using available resources in the most economical way while maintaining the quality and the consistency of the product. An essential prerequisite for process control is having appropriate instruments to measure critical process parameters. Although conventional sensors could measure most process parameters, it is not easy to measure the quality of extruded products on-line. However, it is possible to predict the quality of the product from process measurements using a 'soft sensor'. One of the main problems facing the food industry is that it has to cope with biological raw materials that are inherently variable in quality. Raw material variability is one of the biggest disturbances that are introduced to an extruder and even an experienced process operator may not be able to take corrective action quickly. This would result in the production of out of specification product and consequent financial losses.

It is recommended that a well-designed process monitoring system be implemented before design and development of a control system. A process monitoring system will increase the understanding of the extruder operation and enable operators to control manually the extruder effectively. Furthermore, a process monitoring system could be linked to a quality assurance system.

Designing an automatic control system for an extruder is difficult because it has multiple inputs and outputs. Furthermore, these are non-linear and interact with each other. However, recent developments in control engineering have made it possible to control automatically a pilot scale extruder by using a combination of a soft sensor and a model predictive controller.

The development of a robust, reliable and affordable control system would have a significant financial benefit in a production environment. It is likely that recent advances in sensor technology, soft sensors and advanced model-based control systems would result in the commercial development of a control system for an extruder. It is suggested that a multi-disciplinary team of scientists and engineers develops such a system because it is essential to understand the complex interaction between the raw materials, the process and the product in order to develop a successful control system for an extruder.

It must be pointed out that an advanced control system on its own will not solve the problem of producing inferior quality products. The extrusion process must be regarded as part of an integrated system involving pre- and post-extrusion operations. It is therefore necessary to control all these operations in an integrated manner in order to produce good quality product. Furthermore, other control systems such as Good Manufacturing Practice (GMP) and Hazard Analysis Critical Control Points (HACCP) should also be implemented as part of an overall quality control system.

5.8 Sources of further information and advice

The following research and academic institutes are active in food process control and extrusion process control.

Research Institutes
Food Science Australia *(www.foodscience.afisc.csiro.au)*

Academic Institutes
Centre for Integrated Dynamics and Control, University of Newcastle, Australia *(http://murray.newcastle.edu.au/cidac/)*
University of Sydney *(www.chem.eng.usyd.edu.au)*
Imperial College of Science and Technology, London *(www.ps.ic.ac.uk)*
University College, London *(www.chemeng.ucl.ac.uk)*
Purdue University, USA *(www.foodsci.purdue.edu/cifmc)*
University of Missouri, USA *(www.fse.mossouri.edu/befaculty.htm)*
ENSIA, France *(www.ensia.inra.fr)*

5.9 References

1. MARLIN TE, PERKINS JD, BARTON GW and BRISK ML, 'Benefits from process control: results of a joint industry-university study', *J. Proc. Cont*, 1991 **1** March 68.
2. BRISK ML, 'Process control: theories and profits,' *Proceedings of the 12th World Congress of the International Federation of Automatic Control, Sydney*, 1993 **7** 241.
3. WHITE RG, 'Production process controls', *PROC. ASBE 71st Technical Conference*, 1995 107.
4. MORRIS CE, '1997 Survey of manufacturing trends – the future is now', *Food Engineering* March 1997 67.
5. KULSHRESHTHA MK, ZAROR CA and JUKES DJ, 'Automatic control of extrusion: problems and perspectives', *Food Control*, 1991 **2**(2) 80–86.
6. HOFER JM and TAN J, 'Extrudate temperature control with disturbance rejection', *Food Control*, 1993 **4**(1) 17–24.
7. SINGH B and MULVANEY SJ, 'Modeling and process control of twin-screw cooking food extruders', *Journal of Food Engineering*, 1994 **23** 403–28.
8. ONWALATA CI, MULVANEY SJ and HSIEH F, 'System analysis as the basis for control of density of extruded cornmeal', *Food Control*, 1994 **5**(1) 39–48.
9. BREADIE WLP, MOTTRAM DS and GUY RCE, 'Aroma volatiles gencrated during extrusion cooking of maize flour', *J Agricultural and Food Chemistry*, 1998 **46** 1487–97.
10. SOLINA M, WHITFIELD F, JOHNSON B, BAUMGARTNER P, SLEIGH R and CHESSARI C, 'Generation of aromas in extruded products', *10th World Congress of Food Science and Technology*, Sydney, 1999.

11. BARRETT A H and PELEG M, 'Extrudates cell structure–texture relationships', *J Food Sci*, 1992 **57**(5) 1253–6.
12. VAN HECKE E, ALLAF K and BOUVIER J M, 'Texture and structure of crispy puffed food products I: mechanical properties in bending', *J Texture Studies*, 1995 **26** 11–25.
13. ALVAREZ-MARTINEZ L, KONDURY K P and HARPER J M, 'A general model for expansion of extruded products', *J Food Sci*, 1998 **53**(2) 609–15.
14. MOORE D, SANEI A, VAN HECKE E and BOUVIER J M, 'Effect of ingredients on physical/structural properties of extrudates', *J Food Sci*, 1990 **55**(5) 1383–7.
15. QUEGUINER C, DUMAY E, CAVALIER C and CHEFTEL J C, 'Reduction of *streptococcus thermophilus* in a whey protein isolate by low moisture extrusion cooking without loss of functional properties', *Int J Food Sci & Tech*, 1989 **24** 601–12.
16. BULUT S, WAITES W M and MITCHELL J R, 'Effects of combined shear and thermal forces on destruction of *microbacterium lacticum*', *App Env Micro*, 1999 **65**(10) 4464–9.
17. CULHANE J, SCURRAH K, SELLAHEWA J, SZABO L and RAYNOR D, 'Destruction of microorganisms by extrusion – a feasibility study', Internal Report, Food Science Australia, 2000.
18. FRANCIS G, 'An acoustic method for the on-line prediction of quality in food extrusion', Internal Report Food Science Australia, 1999.
19. EVANS A J, HUANG S, CHESSARI C, OSBORNE B G, KOTWAL Z, WESLEY I J and SAUNDERS L, 'The evaluation of NIR spectroscopy for the on-line monitoring of starch transformations in an extruder', *10th World Congress of Food Science and Technology*, Sydney, 1999.
20. EVANS A J, HUANG S, OSBORNE B G, KOTWAL Z and WESLEY I J, 'Near infrared on-line measurement of degree of cook in extrusion processing of wheat flour', *J Near Infrared Spectroscopy*, 1999 **7** 77–84.
21. WOOD W and GUTERMAN L, 'New electronic "Tongue" can taste what's in a complex mixture', *New Scientist*, 18 July, 10.
22. HALEY T A, 'System identification and control design of an extrusion cooking and puffing process', *Proceedings of ACOFOP IV*, 1998 (Edited by Skjöldebrand C and Trystram G) Gothenburg, Sweden, 1998.
23. KULSHRESHTHA M K, ZAROR C A and JUKES D J, 'Simulating the performance of a control system for food extruders using model-based set-point adjustment', *Food Control*, 1995 **6**(3) 135–41.
24. LU Q, MULVANEY S J and HUFF H E, 'Model and strategies for computer control of a twin-screw extruder', *Food Control*, 1993 **4**(1) 25–33.
25. MOREIRA R G, SRIVASTAVA A K and GERRISH J B, 'Feedforward control model for a twin-screw extruder', *Food Control*, 1990 **1**(3) 179–84.
26. SCHONAUER S and MOREIRA R, 'Development of a fixed-GPC controller for a food extruder based on product quality attributes. Part I: System identification', *Trans IChemE (Part C)*, 1995 **73** 189–99.
27. SCHONAUER S and MOREIRA R, 'Development of a fixed-GPC controller for

a food extruder based on product quality attributes. Part II: Control development, implementation, and analysis', *Trans IChemE(Part C)*, 1995 **73** 200–10.

28. EERIKAINEN T, ZHU Y and LINKO P, 'Neural networks in extrusion process identification and control', *Food Control*, 1994 **5**(2) 111–19.
29. ALTNER K, 'Modelling and control studies of a pilot scale twin screw extruder', Internal Report, Food Science Australia, 1999.
30. GARCIA CE, PRETT DM and MORARI M, 'Model predictive control: theory and practice – a survey', *Automatica*, 1989 **25** 335–48.
31. *Connoisseur*™, Predictive Control Limited, Northwich, Cheshire, UK.
32. CHESSARI C, CHEN J, SELLAHEWA J, VERSTEEG C, WANG L and KARPIEL E, 'A soft sensor approach for the control of product quality in a cooking extruder', *Eighth International Congress on Engineering and Food*, Mexico, 2000.

6

Extrusion and nutritional quality

M. E. Camire, University of Maine

6.1 Introduction

Extruders offer food scientists a palette of conditions and ingredients from which new foods may be created. Although snack foods were among the first commercially successful extruded foods, today extruders produce many foods of nutritional importance. Specific product categories are described elsewhere in this book, yet in this chapter the numerous factors that influence nutritional quality in all products will be explored. Emphasis will be placed on human foods, but extrusion effects certainly have applicability to animal feeds.

How important are extruded foods in the diet? Comprehensive survey data on extruded food consumption is not yet available. The use of extruded products such as breading further complicates the situation. US snack food consumption increased 200% from 1977 to 1994, and ready-to-eat (RTE) breakfast cereal consumption increased by 60% in the same period (Putnam and Gerrior, 1999). Whole grain consumption in the US is barely one serving per day, far below the recommended three servings. Whole grains can be stabilized by extrusion and transformed into palatable cereals. Substitution of RTE cereal for traditional Finnish breakfast foods lowered fat consumption and serum total cholesterol levels (Kleemola *et al.*, 1999). A new trend is to market sweetened cereals as snacks.

Many consumers perceive snack foods to be unhealthy (Dinkins, 2000). A report from the US Department of Agriculture claiming that snacks provide 20% of total energy for American children also found that snacks provided similar percentages for many key nutrients (USDA, 2000). The same report found that over 40% of children regularly eat RTE breakfast cereals. Similar trends likely exist in other developed nations. Childhood obesity and Type II diabetes are increasing in persons under the age of eighteen years, thus increasing risks for developing heart disease (Chen *et al.*, 2000). An interesting dilemma has arisen: should there be more nutrition education to eat more fruits, vegetables and low-fat grains to stop this trend, or should food companies accommodate children's eating habits and develop low-calorie, low-fat low-glycemic index snacks and confections?

Extrusion offers hope for improving nutrition in less-developed nations. Extrusion can produce shelf-stable foods free from microbial contamination that can be stored in preparation for famines and natural disasters. Simple single-screw extruders are relatively inexpensive and easy to maintain, thus international research projects have focused on these machines as food processors for the needy. External energy supplies may not be necessary, as friction from the rotation of the screw may be sufficient to thoroughly cook the food. Processes can be developed to take advantage of donated foods such as dried milk as well as indigenous crops such as beans, millet, and cassava. Use of local crops may help improve sustainability of operations past immediate crises and reduce reliance on foreign aid. Extruded collets may be easily ground into flour that can be mixed with milk or water to form gruel for infants or weakened individuals. Advantages and limitations of extrusion for this application have been reviewed (Harper and Jansen, 1985).

The ability of extruders to blend diverse ingredients in novel foods can also be exploited in the developing functional foods market. Functional ingredients such as soy and botanicals that are relatively unpalatable alone can be incorporated into new food items. Traditional foods such as rye crispbread can be further enhanced by addition of extra dietary fiber during extrusion. Anti-nutritive compounds can be reduced during extrusion to provide safer and more nutritious foods.

There is no clear trend for effects of extrusion on nutritional quality. Much of the basic research on the stability of micronutrients during extrusion was conducted in the 1970s and 1980s. More recent research has focused on non-nutritive healthful food ingredients and novel ingredient blends. Extrusion parameters affecting nutritional quality are summarized in Table 6.1.

High moisture (>40%) extrusion is gaining new applications such as enzymatic conversion of starch for glucose syrup and for production of

Table 6.1 Extrusion parameters affecting nutritional quality

- Feed material composition
- Prior processing history of feed materials
- Water content
- Material feed rate
- Screw speed
- Screw configuration
- Barrel temperature
- Injection
- Die configuration

texturized protein foods and cheese-like products (Akdogan, 1999). Researchers at Cornell University have developed an extrusion process using supercritical carbon dioxide puffing (Sokhey *et al.*, 1996). Both types of extrusion operate at relatively low temperatures with low shear, with the latter designed for puffed products. Nutritional evaluation of these processes has not been published, but it is reasonable to conclude that heat-sensitive ingredients fare better with these methods than with conventional extrusion cooking.

6.2 Macronutrients

6.2.1 Carbohydrates

Starchy tubers and grains provide important energy and satiation in most diets. Sugars provide sweetness and are involved in numerous chemical reactions during extrusion. Control of carbohydrates during extrusion is critical for product nutritional and sensory quality. Extrusion conditions and feed materials must be selected carefully to produce desired results. For example, a weaning food should be highly digestible, yet a snack for obese adults should contain little digestible material. A comparison of how a single formulation may be altered to achieve both goals is shown in Table 6.2.

Humans and other monogastric species cannot easily digest ungelatinized starch. Extrusion cooking is somewhat unique because gelatinization occurs at much lower moisture levels (12–22%) than is necessary in other food operations. Processing conditions that increase temperature, shear, and pressure tend to increase the rate of gelatinization. The presence of other food compounds, particularly lipids, sucrose, dietary fiber and salts, also affects gelatinization (Jin *et al.*, 1994). Complete gelatinization may not occur, but digestibility is improved nonetheless (Wang *et al.*, 1993a).

In a sense, extrusion may pre-digest starch. Branches on amylopectin molecules are easily sheared off in the barrel. Reduction in molecular weight for both amylose and amylopectin molecules have been documented. Politz and co-workers (1994b) found that larger corn amylopectin molecules were subjected to the greatest molecular weight reduction. In a related study, wheat flour starch showed greater starch degradation after extrusion, and higher die temperature

Table 6.2 Modification of extrusion conditions to yield products with varying starch digestibility from corn meal

High digestibility weaning food	High-resistant starch, low-calorie food
Feed moisture > 18%	Low feed moisture
Mass temperature > 120°C	Low barrel temperature
High-shear screw configuration with added reverse elements	Kneading blocks to increase residence time
Added vitamins, minerals	Added citric acid, < 1 %

(185°C) and feed moisture (20%) helped to maintain molecular weight (Politz *et al.*, 1994a). Screw configuration can be designed to minimize or maximize starch breakdown (Gautam and Choudhoury, 1999).

Rapidly-digested starch triggers rapid rise in blood sugar and insulin levels after meals. These increases may lead to insulin insensitivity and Type II, or adult-onset, diabetes. The rise in blood glucose after eating is often measured as the glycemic index (GI), with glucose or white bread used as an arbitrary control with a value of 100. High amylose rice extruded into noodles had lower starch digestibility and reduced GI in human volunteers (Panlasigui *et al.*, 1992). Higher GI values were found for persons with diabetes.

Extrusion conditions can be manipulated to produce digestion-resistant starch (RS) by several mechanisms. As branches are removed from amylopectin molecules, they could react with other carbohydrates in novel linkages that cannot be digested by our enzymes. Theander and Westerlund (1987) reported such transglycosidation in extruded wheat flour, but analytical limitations have made it difficult for other researchers to corroborate this finding. Politz *et al.* (1994a) did not observe any differences in 2,3-glucose linkages in extruded wheat flour after methylation analysis. A variation on this process exists in a patent for producing as much as 30% resistant starch in which high amylose starch is reacted with pullulanase, then extruded (Chiu *et al.*, 1994).

Other options for increasing RS have been published. In a series of three experiments, Unlu and Faller (1998) demonstrated that adding certain forms of starch or citric acid to corn meal prior to extrusion modified resistant starch plus dietary fiber. Addition of 30% corn, potato or wheat starch did not increase RS values. RS and fiber values more than doubled when 7.5% citric acid was mixed with cornmeal, and 30% high-amylose cornstarch with 5 or 7.5% citric acid resulted in values of 14%, compared with slightly more than 2% in 100% cornmeal. The authors speculated that polydextrose may have been formed within the extruder. Cost and sensory acceptability were not determined. Oligosaccharides and polydextrose were formed from glucose-citric acid mixtures extruded at different barrel temperatures (Hwang *et al.*, 1998). Yields of polymers increased with temperature; 93.7% yield occurred at 200°C.

Added dietary fiber also affects digestibility. Longer cellulose fibers added to cornstarch decreased starch solubility (Chinnaswamy and Hanna, 1991). On the other hand, addition of 20% protein with removal of insoluble dietary fiber from wheat flour resulted in pasta with significantly delayed dextrin release under *in vitro* digestion conditions (Fardet *et al.*, 1999). Microscopic evaluation indicated that fiber removal facilitated starch-protein interactions that may have increased enzyme-resistance.

Amylose-lipid complex formation can also reduce starch digestibility. Monoglycerides and free fatty acids are more likely to form complexes than are triglycerides when added to high-amylose starch (Bhatnagar and Hanna, 1994a). Stearic acid mixed with normal corn starch with 25% amylose and extruded at low feed moisture (19%) and low barrel temperature (110–140°C) contained the most starch-lipid complex (Bhatnagar and Hanna, 1994b). Conditions of high viscosity and longer residence time favored complex formation in small extruders.

The adverse nutritional consequences of easily-digested starch include increased risks for dental caries and rapid rise in blood glucose levels after eating. The smaller starch fragments formed during extrusion may be sticky and thus could adhere to teeth. Toothpack, the amount of material retained on teeth after eating extruded foods, can be used as an indication of the severity of processing. Dental plaque bacteria rapidly ferment dextrins. White wheat flour extruded under 'mild' and 'severe' conditions caused drops in dental plaque pH comparable to glucose solution (Björck *et al.*, 1984).

The fate of sugars during extrusion cooking cannot be overlooked. Biscuits enriched with protein underwent sucrose hydrolysis during extrusion, with sucrose losses of 2–20% (Noguchi and Cheftel, 1983). Reducing sugars are presumably lost during Maillard reactions with proteins. Sucrose, raffinose and stachyose decreased significantly in extruded high-starch fractions of pinto beans (Borejszo and Khan, 1992). Extruded snacks based on corn and soy contained lower levels of both stachyose and raffinose compared to unextruded soy grits and flour, but values were not corrected for the 50–60% corn present (Omueti and Morton, 1996). The destruction of these flatulence-causing oligosaccharides may improve consumer acceptance of extruded legume products.

6.2.2 Proteins and amino acids

Extrusion improves protein digestibility via denaturation, which exposes enzyme-access sites. Most proteins such as enzymes and enzyme inhibitors lose activity due to denaturation. The extent of denaturation is typically assessed as change in protein solubility in water or aqueous solutions. These changes are more pronounced under high shear extrusion conditions (Della Valle *et al.*, 1994), although mass temperature and moisture are also important influences. For example, wheat protein solubility is reduced even at the relatively low process temperatures used in pasta making (Ummadi *et al.*, 1995a). Arêas (1992) and Camire (1991) have written reviews on protein extrusion.

Since most extruded foods are not high in protein, nutritional evaluations of extruded feeds, weaning foods and other specialized products have been emphasized. High barrel temperatures and low moistures promote Maillard reactions during extrusion. Reducing sugars, including those formed during shear of starch and sucrose, can react with lysine, thereby lowering protein nutritional value. Lysine is the limiting essential amino acid in cereals, and further depletion of this nutrient can impair growth in children and young animals. Blends of cornmeal, full-fat soy flakes and soy isolates or concentrates produced nutritious ingredients suitable for reconstitution as porridge or gruel with good retention of lysine (Konstance *et al.*, 1998). Acidic pH increased Maillard reactions in a model system consisting of wheat starch, glucose and lysine (Bates *et al.*, 1994).

Soy protein has been identified as a cholesterol-lowering food ingredient. Extrusion texturization of soy isolate did not reduce its effects on rat serum cholesterol, excretion of cholesterol and other steroids in feces, or protein nutrition compared with non-extruded soy (Fukui *et al.*, 1993).

6.2.3 Lipids

Although lipids serve as a concentrated form of energy, excess dietary lipid consumption is associated with chronic illnesses such as heart disease, cancer, and obesity. Generally foods containing less than ten per cent lipids are extruded because greater quantities of lipids reduce slip within the extruder barrel, making extrusion difficult, particularly for expanded products. Many extruded snack foods are fried after extrusion to remove moisture and modify texture and flavor. Response surface methodology was used to determine which extrusion parameters would limit oil absorption by a multigrain chip during frying (Osman *et al.*, 2000). Oil content ranged from 20 to 35% as compared to a commercial product that has only 21% oil, suggesting that further improvements are necessary. Extrusion can be used to aid oil extraction since oil is freed during the cooking and shearing operations (Nelson *et al.*, 1987).

Such obvious loss of lipids was first used to explain a mass balance dilemma in extrusion: why do extruded foods appear to contain lower lipid levels? The answer lay in formation of starch-lipid complexes resistant to some lipid extraction techniques. Lipid recovery is improved when extruded foods are first digested with acid or amylase, then extracted with an organic solvent such as ether. Total fat was not significantly changed in extruded whole wheat, but only half of the ether-extractable lipids in extruded was detected (Wang *et al.*, 1993b). After extrusion, wheat bran, which is lower in starch content than whole wheat, had more free lipids. Cornmeal extruded at 50–60°C or 85–90°C contained over 75% bound lipids, but extrusion at 120–125°C only bound 70% of the lipids (Guzman *et al.*, 1992).

Despite interest in health benefits of omega-3 fatty acids, only one study has been published on the stability of these highly unsaturated lipids. Both docosahexaenoic (DHA) and eicosapentaenoic (EPA) acids were retained in chum salmon muscle extruded with ten per cent wheat flour (Suzuki *et al.*, 1988). Another nutritional issue is the safety of *trans* fatty acids. Maga (1978) found that extrusion of corn and soy resulted in formation of only 1.5% *trans* fatty acids.

Lipid oxidation is a major cause of loss of nutritional and sensory quality in foods and feeds. Although we suspect that lipid oxidation does not occur during extrusion due to brief residence time, lipid oxidation can occur during storage. Artz *et al.* (1992) have reviewed factors affecting lipid oxidation in extruded foods. Screw wear results in higher concentrations of pro-oxidant minerals. Semwal *et al.* (1994) found that iron and peroxide values were higher in extruded rice and dhal compared to dried products. Another factor favoring oxidation is the formation of air cells in expanded products, leading to increased

Table 6.3 Antioxidant schemes for extruded foods

Antioxidant treatment	Advantages	Limitations
Butylated hydroxyanisole (BHA); butylated hydroxytoluene (BHT)	Disperse well in dry feed	Volatility results in low levels in final product
Tocopherols	Varying vitamin E activity; appeal for consumers wanting natural products	Not heat-stable; common forms are difficult to meter in dry feed or water
Rosemary extract	Natural product	Flavor incompatibility (does not apply to odor-free forms)
Phenolic acids	Natural; cost may be low if food by-products are the source	Darker color due to complex formation; astringent or bitter flavor
Ascorbic acid	Natural; vitamin C activity; enhances iron bioavailability	Not heat-stable; high levels impart acidic flavor
Nitrogen flush in packaging	No direct additives	Some air is trapped within the expanded food, so oxidation is possible; cost of gas and packaging

surface area. However, lipolytic enzymes and other enzymes that promote oxidation may be inactivated during extrusion, and starch-lipid complexes formed in the barrel may be more resistant to oxidation. Packaging under nitrogen or vacuum in opaque containers may further protect extruded foods. The relative effectiveness of different antioxidants is discussed in Table 6.3.

6.2.4 Dietary fiber

Fiber is a term used to describe many food components. In 1999 the American Association of Cereal Chemists coined the following description of dietary fiber:

> Dietary fiber is the edible parts of plants or analogous carbohydrates that are resistant to digestion and absorption in the human small intestine with complete or partial fermentation in the large intestine. Dietary fiber includes polysaccharides, oligosaccharides, lignin, and associated plant substances. Dietary fibers promote beneficial physiological effects including laxation, and/or blood cholesterol attenuation, and/or blood glucose attenuation.

A major difficulty in interpreting research involving fiber and extrusion is the variety of analytical methods used to quantitate and characterize different fiber components. For example, the AOAC total dietary fiber method required for US nutritional labeling does not measure compounds that are soluble in 80% aqueous ethanol. Measurement of total dietary fiber for food labeling does not detect changes in fiber solubility induced by extrusion. As with starch, fragments of larger molecules may be sheared off during extrusion. These smaller molecules may be water-soluble. On the other hand, fragments could unite to form large insoluble complexes or Maillard compounds that may be analyzed as lignin. Such physicochemical changes may influence profoundly the health benefits of the extruded foods. For example, soluble forms of dietary fiber are associated with reduced risks for heart disease.

Extrusion did not affect uronic acids (components of pectin) but insoluble nonstarch polysaccharides (NSP) increased in oatmeal and potato peels (Camire and Flint, 1991). Soluble NSP was higher in extruded oatmeal and potato peels, and corn meal fiber was unaffected by extrusion. Beans (*Phaseolus vulgaris* L) subjected to conditions making them hard-to-cook, were extruded under various conditions in order to make them more functional (Martín-Cabrejas *et al.*, 1999). Total fiber values were unaffected by extrusion but insoluble fiber decreased when extruded at 25% moisture content. Soluble increased in those samples, and especially in a sample processed at 30% moisture and 180°C.

How does this redistribution of molecules affect nutritional quality? Conflicting findings have been reported. These studies are summarized in Table 6.4. The viscosity of aqueous suspensions of extruded wheat, oats and barley were higher than unprocessed grains (Wang and Klopfenstein, 1993). Viscous gums and other soluble fibers may reduce cholesterol levels by trapping bile acids; increased excretion of bile eventually depletes body stores of cholesterol, which are tapped to synthesize new bile acids.

Soluble forms of fiber such as those found in fruit and gums form gels in the small intestine. The increased viscosity is believed to retard the absorption of

Table 6.4 Nutritional effects of dietary fiber extrusion

Food source	Dietary fiber change	Nutrition assay	Health effect	Reference
Extruded wheat, barley with husks, or oats with husks	↑ SDF	6-wk rat feeding study	↓ cholesterol in rats fed extruded grains versus raw grains or casein control	Wang and Klopfenstein, 1993
Potato peels	↓ SDF at lower barrel temperature	*In vitro* bile acid binding	↓ binding could lower serum cholesterol	Camire *et al.*, 1993
Wheat cereal with added guar gum		Human glucose tolerance test	↓ serum glucose postprandial compared to low-fiber cereal	Fairchild *et al.*, 1996
Extruded rice, oat, corn and wheat brans	No change	Hamster feeding study	Extrusion did not affect cholesterol-lowering properties	Kahlon, *et al.*, 1998

glucose, preventing spikes in post-prandial serum glucose levels. Increased levels of soluble fiber in citrus peels after extrusion were correlated with increased *in vitro* viscosity (Gourgue *et al.*, 1994). However, starch digestion and diffusion of glucose were not affected by extrusion. Extrusion reduced sugar beet pectin and hemicellulose molecular weight and viscosity, but water solubility increased 16.6% to 47.5% (Ralet *et al.*, 1991). Extrusion of guar gum in a wheat flake cereal did not impair the gum's ability to lower postprandial blood insulin and glucose levels in healthy adults (Fairchild *et al.*, 1996). However, 92 g/day extruded dry white beans fed in baked goods did not lower serum lipoproteins in middle-aged men with hyperlipidemia (Oosthuizen *et al.*, 2000). Extruded beans, suggesting a potential risk reduction for cardiovascular disease, decreased plasminogen activator factor inhibitor 1 levels.

Beta-glucans in oats and barley are believed to be responsible for the cholesterol-lowering properties of those grains. Solubility of β-glucans in regular (Phoenix) and waxy (Candle) barley cultivars increased after extrusion at four barrel temperatures and three different moisture contents (Gaosong and Vasanthan, 2000).

Although insoluble forms of dietary fiber are thought to maintain colonic health, the value of fiber in preventing colon cancer has been questioned (Fuchs *et al.*, 1999; Alberts *et al.*, 2000). One theory is that dietary fiber provides protection against colorectal cancer by binding dietary carcinogens. Potato peels extruded at 110°C barrel temperature and 30% feed moisture significantly reduced binding of the polycyclic aromatic hydrocarbon benzo[a]pyrene (Camire *et al.*, 1995a). Sixteen extruded commercial cereals bound at least 40% of benzo[a]pyrene, regardless of fiber content (Camire *et al.*, 1995b).

6.3 Vitamins

Killeit reviewed vitamin retention in extruded foods in 1994. Although fortification of extruded foods with micronutrients is popular, little research has examined the interaction of extrusion conditions and nutrients. Concerns of reduced vitamin levels prompt some manufacturers to apply vitamins post-extrusion as a spray. More recent research has focused on vitamin stability in feeds. Fat-coated ascorbic acid, menadione, pyridoxine and folic acid were retained better than crystalline forms in extruded fish feed (Marchetti *et al.*, 1999).

6.3.1 Vitamin A and the carotenoids

Vitamin A deficiency is a major cause of blindness in many less-developed nations, and the vitamin is important for healthy immune system function. Unfortunately oxygen and heat destroy vitamin A and related carotenoids. Beta-carotene is an antioxidant that is a vitamin A precursor. Beta-carotene is added to foods to make them more orange in color, but it is unstable when heated.

Increasing barrel temperatures from 125 to 200ºC resulted in more than 50% destruction of all *trans* β-carotene in wheat flour 50% (Guzman-Tello and Cheftel, 1990). Fifteen colored degradation products of all *trans* β-carotene dispersed in cornstarch were recovered after twin-screw extrusion (Marty and Berset, 1988).

6.3.2 Other lipid-soluble vitamins

Vitamins D and K are fairly stable during food processing, and they are not commonly used in extruded human foods. Vitamin E and related tocopherols, however, perform as both vitamin and antioxidant. Tocopherol and retinyl palmitate decreased in puffed snacks containing either fish or partially-defatted peanut flour (Suknark, 1998). Rice bran tocopherol decreased with increasing extrusion temperature, and bran extruded at 120–140ºC lost more tocopherols over a year's storage than did bran extruded at 110ºC (Shin *et al.*, 1997). Less than 20% of vitamin E was retained in extruded and drum-dried wheat flour (Wennermark, 1993).

6.3.3 Ascorbic acid

Ascorbic acid (vitamin C) is lost in the presence of heat and oxidation. This vitamin decreased in wheat flour when extruded at higher barrel temperatures at fairly low moisture (10%) (Andersson and Hedlund, 1990). Blueberry concentrate appeared to protect 1% added vitamin C in a extruded breakfast cereal compared to a product containing just corn, sucrose, and ascorbic acid (Chaovanalikit, 1999). When ascorbic acid was added to cassava starch to increase starch conversion, retention of over 50% occurred at levels of 0.4–1.0% addition (Sriburi and Hill, 2000).

6.3.4 The B vitamins

Grains must be enriched with B vitamins in the United States. Thiamine is the water-soluble vitamin most susceptible to thermal processing. Thiamine destruction in extruded wheat flour is a first-order reaction (Guzman-Tello and Cheftel, 1987). Killeit (1994) summarized thiamine losses as ranging from 5 to 100%. Thiamine retention in potato flakes decreased under extrusion conditions of lower moisture and higher barrel temperature; sulfites in the potato flakes may have also contributed to vitamin destruction (Maga and Sizer, 1978). Large losses of thiamine occurred when no water was added during extrusion, but riboflavin (B_2) and niacin were not affected (Andersson and Hedlund, 1990). Using low-cost single-screw extruders, Lorenz and Jansen (1980) found retention of over 90% for thiamin, riboflavin, vitamin B_6 and folic acid in corn-soy blends processed at 171ºC.

Folate is the most recent vitamin to be required for fortification and enrichment. The term folate is used to describe a family of related pteroylpolyglutamates and folic acid, a synthetic vitamin. Consumption of adequate folate by pregnant women and women of child-bearing age is recommended to prevent neural tube birth defects. Folic acid has superior bioavailability to folates found naturally in foods (Institute of Medicine, 1998), thus stability of folic acid in extruded foods should be evaluated.

6.4 Minerals

6.4.1 Screw wear

High-fiber foods may abrade the interior of the extruder barrel and screws, resulting in increased mineral content. Potato peels extruded under higher temperature had as much as 38% more total iron after extrusion (Camire *et al.*, 1993). Extruded corn, which is fairly low in fiber, showed no difference in total, elemental, or soluble iron, even in the presence of antioxidant additives (Camire and Dougherty, 1998). Iron content in extruded potato flakes increased with barrel temperature (Maga and Sizer, 1978). Screw wear iron had high bioavailability in rats fed extruded corn and potato (Fairweather-Tait *et al.*, 1987). Extrusion and any resulting changes in mineral content did not reduce utilization of iron and zinc from wheat bran and wheat in adult human volunteers (Fairweather-Tait *et al.*, 1989).

The solubility of iron under conditions similar to digestion and subsequent ability to dialyze across a membrane is used to assess bioavailability. Extrusion slightly increased iron availability in corn snacks (Hazell and Johnson, 1989). High-shear extrusion reduced dialyzable iron compared to low-shear extrusion of navy beans, lentils, chickpeas and cowpeas (Ummadi *et al.*, 1995b). Weaning foods based on pearl millet, cowpea and peanut had higher iron availability and protein digestibility than did similar foods prepared by roasting, however none of the blends provided adequate iron to meet infant needs (Cisse *et al.*, 1998). Extrusion did not compromise the zinc bioavailability of 85:15 blends of semolina and soy protein concentrate (Kang, 1996).

6.4.2 Other issues involving minerals

Loss of mineral bioavailability may occur in foods containing high levels of dietary fiber and phytate. Gualberto and colleagues (1997) found that varying screw speed had no effect on phytate retention in wheat, rice and oat brans, but insoluble fiber decreased in rice and oat bran after extrusion. When phytate was removed from these samples, extruded rice and oat brans bound more calcium and zinc *in vitro*, but not copper (Bergman *et al.*, 1997). Similar results were observed with a high-fiber cereal product fed to seven persons with ileostomies; while dietary fiber and phytate values were not affected by extrusion, mineral availability was reduced (Sandberg *et al.*, 1986; Kivistö *et al.*, 1986). Extrusion reduced phytate levels in wheat flour (Fairweather-Tait et al, 1989). Inactivation of phytases during extrusion in these studies may partly explain these findings.

Although phytic acid was lower under all processing conditions, total phytate was not affected. Legume phytate was also not affected by extrusion (Lombardi-Boccia *et al.*, 1991; Ummadi *et al.*, 1995b).

Interest in nutrient fortification has led to addition of minerals to extruded foods, particularly cereals. Added calcium hydroxide (0.15–0.35%) decreased expansion and increased lightness in color of cornmeal extrudates (Martínez-Bustos *et al.*, 1998), but bioavailability of added calcium after extrusion has not been reported. Certain iron salts react with phenolics to form unattractive dark colors. Kapanidis and Lee (1996) recommended the use of ferrous sulfate heptahydrate in a simulated rice product for maintaining light color. Mineral fortification, in light of screw wear, should be evaluated for bioavailability of key nutrients.

6.5 Non-nutrient healthful components of foods

6.5.1 Phenolic compounds

Phenolic compounds in plants protect against oxidation, disease, and predation. These compounds, including the large flavonoid family, are the focus of numerous studies to elucidate their role in human health. In potato peels, total free phenolics, of which chlorogenic acid predominates, were lower post-extrusion (unpublished data, Camire and Dougherty). Higher barrel temperature and feed moisture protected free phenolics.

The red and blue anthocyanin pigments provide attractive colors and are believed to serve as antioxidants that protect vision and cardiovascular health (Camire, 2000). Blueberry anthocyanins were significantly reduced by extrusion and by ascorbic acid in breakfast cereals containing cornmeal and sucrose (Chaovanalikit, 1999). Polymerization and browning may also have contributed to anthocyanin losses.

6.5.2 Antinutrients

Extrusion cooking destroys many natural toxins and antinutrients (Table 6.5), thereby improving safety and digestibility of the foods. Enzyme inhibitors, hormone-like compounds, saponins and other compounds may impair growth in children but may protect adults against chronic diseases. Compounds such as allergens and mycotoxins that are more resistant to heat and shear may be susceptible to extrusion in combination chemical treatments.

Glucosinolates are found in many commercially-important *Brassica* species, and may have a role in cancer prevention (Van Poppel *et al.*, 1999). Extrusion alone likely has little effect on retention of glucosinolates (Fenwick *et al.*, 1986). Canola total glucosinolates were reduced by added ammonia during extrusion (Darroch *et al.*, 1990). Although extrusion with ammonium carbonate did not completely destroy glucosinolates in rapeseed meal, the process did improve nutritional parameters in rats fed the extruded versus unprocessed rapeseed meal (Barrett *et al.*, 1997).

Table 6.5 Antinutrients and toxins affected by extrusion

Compound	Factors favoring reduction
Allergens	Increased shear; added starch
Glucosinolates	Added ammonia
Glycoalkaloids	Added thiamine
Gossypol	Higher feed moisture
Mycotoxins	Increased mixing, lower temperatures; added amine sources
Protease inhibitors	Higher extrusion temperatures

Soy isoflavones have estrogenic activity, and thus may protect post-menopausal women from osteoporosis and heart disease, while men may receive protection against prostate and other testosterone-dependent cancers. Okara, a by-product of tofu manufacture, was mixed with wheat flour and evaluated for retention of isoflavones (Rinaldi *et al.*, 2000). Two barrel temperatures and screw configurations were tested. The aglycone genistein significantly decreased under all extrusion conditions, and glucosides of daidzin and genistin increased, presumably at the expense of acetyl and malonyl forms. Total isoflavone values were significantly lower in 40% okara samples extruded at high temperature.

In blends of 20% soy protein concentrate with cornmeal, increasing barrel temperature caused decarboxylation of isoflavones, leading to increased proportions of acetyl derivatives (Mahungu *et al.*, 1999). Total isoflavones also decreased in the soy-corn blends. In a related study, although the content of the biologically-active aglycones did not change with extrusion, extruded corn-soy blends were less effective in preventing proliferation of breast cancer cells *in vitro* (Singletary *et al.*, 2000). Optimization of extrusion conditions to retain health benefits of soy products is clearly needed.

6.6 Future trends

The possibilities are endless for further experimentation in extrusion and nutrition. Very little has been published on the effects of extrusion on phytochemicals and other healthful food components. A list of as-yet unstudied (or unpublished) compounds is given in Table 6.6. Improved chemical and immunoassay methods will undoubtedly facilitate research in this area. As mentioned early in this chapter, evaluations of nutrient retention by either high-moisture extrusion or by supercritical fluid extrusion have yet to be published.

Relatively few universities possess extruders, and those that do typically own small models that are inexpensive to acquire and operate. Improved understanding of scale-up issues in extrusion is necessary for valid interpretation of studies conducted using laboratory-scale and pilot plant extruders. Evaluation of the effects of extrusion is time consuming compared to the rapid production

Table 6.6 Food compounds for which extrusion effects are not well-documented

- Caffeine and other stimulants
- Flavonoids
- Inulin
- Lycopene and related carotenoids
- Omega-3 fatty acids
- Proanthocyanidins
- Resveratrol
- Tannins

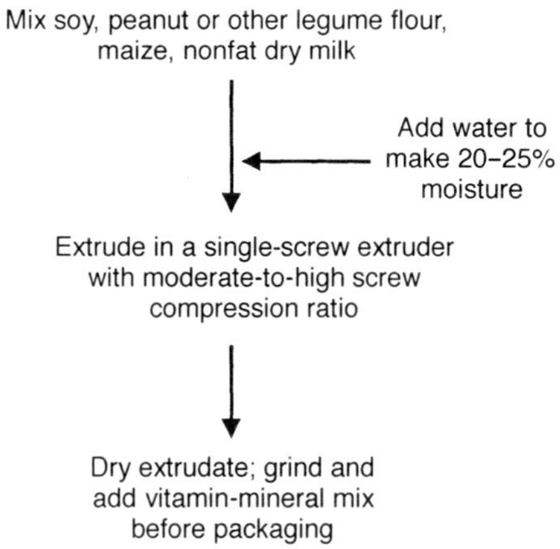

Fig. 6.1 Possible scheme for development of a nutritious food via extrusion cooking.

of extrudates. Long-term animal and feeding studies are especially tedious and costly, yet essential for demonstrating safety and efficacy of extruded foods.

How should product development of healthy extruded foods proceed? Figure 6.1 illustrates a scheme for development of a nutritious food for less-developed nations, while Fig. 6.2 describes considerations for creating a new functional food via extrusion. Integration of nutrition, food science and engineering will be essential for the success of any new product, yet extrusion holds its own unique challenges. The 21st century holds many possibilities for improving human nutrition and extruders will continue to be important tools towards that goal.

6.7 Sources of further information and advice

While no one publication is devoted to food extrusion, several publications, such as *Cereal Chemistry*, *Journal of Agricultural and Food Chemistry*, *Journal of Cereal Science*, *Journal of Food Engineering*, and *Journal of Food Science*, regularly feature extrusion research papers. Chemical and nutritional changes in extruded foods have been the subject of review articles as well (Björck and Asp,

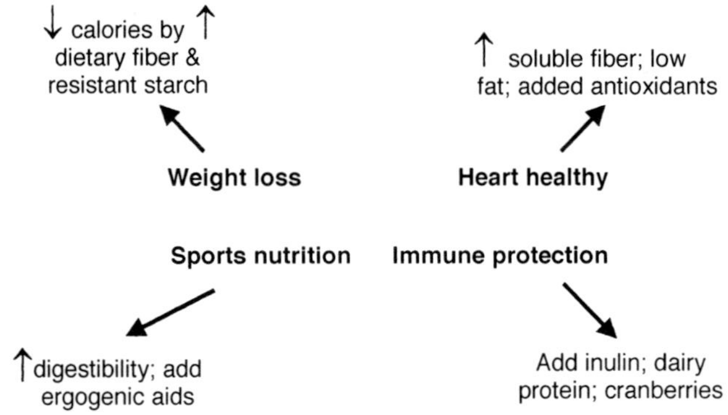

Fig. 6.2 Possible scheme for development of an extruded functional food.

1983; Camire, 1998; Camire *et al.*, 1990; Cheftel, 1986; de la Gueriviere *et al.*, 1985). A handful of books have been published on extrusion cooking (Frame, 1994; Harper, 1981; Hayakawa, 1992; Kokini *et al.*, 1992; Mercier *et al.*, 1989; O'Connor, 1987; Riaz, 2000).

6.8 References and further reading

AKDOGAN H (1999), 'High moisture food extrusion', *Intl J Food Sci Technol*, 34(3), 195–207.

ALBERTS D S, MARTINEZ M E, ROE D J, GUILLEN-RODRIGUEZ J M, MARSHALL J R, VAN LEEUWEN J B, REID M E, RITENBAUGH C, VARGAS P A, BHATTACHARYYA A B, EARNEST D J and SAMPLINER R E (2000), 'Lack of effect of a high-fiber cereal supplement on the recurrence of colorectal adenomas. Phoenix Colon Cancer Prevention Physicians' Network', *N Engl J Med*, 342(16), 1156–62.

ALONSO R, GRANT G, DEWET P and MARZO F (2000), 'Nutritional assessment in vitro and in vivo of raw and extruded peas (*Pisum sativum* L.)', *J Agric Food Chem*, 48(6), 2286–90.

ANDERSSON Y and HEDLUND B (1991), 'Extruded whey flour: correlation between processing and product quality parameters', *Food Quality and Preference*, 2(3), 201–16.

ARÊAS J A G (1992), 'Extrusion of food proteins', *Crit Rev Food Sci Nutr*, 32(4), 365–92.

ARTZ W E, RAO S K and SAUER R M (1992), 'Lipid oxidation in extruded products during storage as affected by extrusion temperature and selected antioxidants', in Kokini J L, Ho C-T, and Karwe M V, *Food Extrusion Science and Technology*, New York, Marcel Dekker, Inc.

BARRETT J E, KLOPFENSTEIN C F and LEIPOLD H W (1997) 'Detoxification of rapeseed meal by extrusion with an added basic salt', *Cereal Chem*, 74(2), 168–70.

BATES L, AMES J M and MACDOUGALL D B (1994), 'The use of a reaction cell to model the development and control of colour in extrusion cooked foods', *Lebensm-Wiss u-Technol*, 27(4), 375–9.

BERGMAN C J, GUALBERTO D G and WEBER C W (1997), 'Mineral binding capacity of dephytinized insoluble fiber from extruded wheat, oat and rice brans', *Plant Foods Hum Nutr*, 51(4), 295–310.

BHATNAGAR S and HANNA M A (1994a), 'Amylose-lipid complex formation during single- screw extrusion of various corn starches', *Cereal Chem*, 71(6), 582–7.

BHATNAGAR S and HANNA M A (1994b), 'Extrusion processing conditions for amylose-lipid complexing', *Cereal Chem*, 71(6), 587–93.

BJÖRCK I and ASP N-G (1983), 'The effects of extrusion cooking on nutritional value – a literature review', *J Food Engin*, 2(4), 281–308.

BJÖRCK I, ASP N-G, BIRKHED D and LUNDQUIST I (1984), 'Effects of processing on availability of starch for digestion *in vitro* and *in vivo*; I. extrusion cooking of wheat flours and starch', *J Cereal Sci*, 2(2), 91–103.

BOREJSZO Z and KHAN K (1992), 'Reduction of flatulence-causing sugars by high temperature extrusion of pinto bean high starch fractions', *J Food Sci*, 57(3), 771–2 and 777.

CAMIRE M E (1991), 'Protein functionality modification by extrusion cooking', *J Am Oil Chem Soc*, 68(3), 200–5.

CAMIRE M E (1998), 'Chemical changes during extrusion cooking. process-induced chemical changes in food', in Shahidi F, Ho C-T, and van Chuyen H, *Process-induced Chemical Changes in Food*, New York, Plenum Press.

CAMIRE M E (2000), 'Bilberries and blueberries as functional foods and pharmaceuticals', in Mazza G and Oomah D B, *Functional Foods: Herbs, Botanicals and Teas*, Lancaster, PA, Technomic Press.

CAMIRE M E and DOUGHERTY M P (1998), 'Added phenolic compounds enhance lipid stability in extruded corn', *J Food Sci*, 63(4), 516–18.

CAMIRE M E and FLINT S L (1991), 'Thermal processing effects on dietary fiber composition and hydration capacity in corn meal, oat meal and potato peels', *Cereal Chem*, 68(6), 645–7.

CAMIRE M E, CAMIRE A and KRUMHAR K (1990), 'Chemical and nutritional changes in foods during extrusion', *Crit Rev Food Sci Nutr*, 29(1), 35–57.

CAMIRE M E, ZHAO J and VIOLETTE D A (1993), '*In vitro* binding of bile acids by extruded potato peels', *J Agric Food Chem*, 41(12), 2391–4.

CAMIRE M E, ZHAO J, DOUGHERTY M P and BUSHWAY R J (1995a), '*In vitro* binding of benzo(a)pyrene by extruded potato peels', *J Agric Food Chem*, 43(4), 970—3.

CAMIRE M E, ZHAO J, DOUGHERTY M P and BUSHWAY R J (1995b), '*In vitro* binding of benzo(a)pyrene by ready-to-eat breakfast cereals', *Cereal Foods World*, 40(6), 447–50.

CHAOVANALIKIT A (1999), 'Anthocyanin stability during extrusion cooking', M S Thesis, Univ. of Maine, Orono ME.

CHEFTEL J C (1986), 'Nutritional effects of extrusion cooking', *Food Chem*, 20, 263–83.

CHEN W, BAO W, BEGUM S, ELKASABANY A, SRINIVASAN S R and BERENSON G S (2000), 'Age-related patterns of the clustering of cardiovascular risk variables of syndrome X from childhood to young adulthood in a population made up of black and white subjects: the Bogalusa Heart Study', *Diabetes*, 49(6), 1042–8.

CHINNASWAMY R and HANNA M A (1991), 'Physicochemical and macromolecular properties of starch-cellulose fiber extrudates', *Food Structure*, 10(3), 229–39.

CHIU C W, HENLEY M and ALTIERI P (1994), 'Process for making amylase resistant starch from high amylose starch', United States Patent. Patent No: 5,281,276. Date of Patent: Jan. 25, 1994.

CISSE D, GUIRO A T, DIAHAM B, SOUANE M, DOUMBOUYA N T and WADE S (1998), 'Effect of food processing on iron availability of African pearl millet weaning foods', *Intl J Food Sci Nutr*, 49(5), 375–81.

DARROCH C S, BELL J M and KEITH M O (1990), 'The effects of moist heat and ammonia on the chemical composition and feeding value of extruded canola screenings for mice', *Can J Anim Sci*, 70, 267–77.

DE LA GUERIVIERE JF, MERCIER C and BAUDET L (1985), 'Incidences de la cuisson-extrusion sur certains parametres nutritionnels de produits alimentaires notamment céréaliers', *Cah Nutr Diet*, 20, 201–10.

DELLA VALLE G, QUILLIEN L and GUEGUEN J (1994), 'Relationships between processing conditions and starch and protein modifications during extrusion-cooking of pea flour', *J Sci Food Agric*, 64(4), 509–17.

DINKINS J (2000), 'Beliefs and attitudes of Americans towards their diet', *Nutrition Insight* 19, 1–2.

FAIRCHILD R M, ELLIS P R, BYRNE A J, LUZIO S D and MIR M A (1996), 'A new breakfast cereal containing guar gum reduces postprandial plasma glucose and insulin concentrations in normal-weight human subjects', *Brit J Nutr*, 46(1), 63–73.

FAIRWEATHER-TAIT S J, PORTWOOD D E, SYMSS L L, EAGLES J and MINSKI M J (1989), 'Iron and zinc absorption in human subjects from a mixed meal of extruded and non-extruded wheat bran flour', *Am J Clin Nutr*, 49(1), 151–5.

FAIRWEATHER-TAIT S J, SYMSS L L, SMITH A C and JOHNSON I T (1987), 'The effect of extrusion cooking on iron absorption from maize and potato', *J. Sci Food Agric*, 39(4), 341–8.

FARDET A, ABECASSIS J, HOEBLER C, BALDWIN P M, BULEON A, BEROT S and BARRY J L (1999), 'Influence of technological modifications of the protein network from pasta on in vitro starch degradation', *J Cereal Sci*, 30(2), 33–145.

FENWICK G R, SPINKS E A, WILKINSON A P, HEANEY R K and LEGOY M A (1986), 'Effect of processing on the antinutrient content of rapeseed', *J Sci Food Agric*, 37(8), 735–41.

FRAME N D *The Technology of Extrusion Cooking*, Glasgow, Blackie Academic & Professional.

FUCHS C S, GIOVANNUCCI E L, COLDITZ G A, HUNTER D J, STAMPFER M J, ROSNER B, SPEIZER F E and WILLETT W C, (1999), 'Dietary fiber and the risk of colorectal cancer and adenoma in women', *N Engl J Med*, 340(3), 169–76.

FUKUI K, AOYAMA T, HASHIMOTO Y and YAMAMOTO T, (1993), 'Effect of extrusion of soy protein isolate on plasma cholesterol level and nutritive value of protein in growing male rats', *J Jap Soc Nutr Food Sci*, 46(3), 211–16.

GAOSONG J and VASANTHAN T, (2000), 'Effect of extrusion cooking on the primary structure and water solubility of ß-glucans from regular and waxy barley', *Cereal Chem*, 77(3), 396–400.

GAUTAM A and CHOUDHOURY G S, (1999), 'Screw configuration effects on starch breakdown during twin screw extrusion of rice flour', *J Food Process Preserv*, 23(4), 355–75.

GOMEZ-ALDAPA C, MARTÍNEZ-BUSTOS F, FIGUEROA CJD and ORDORICA FCA (1999) 'A comparison of the quality of whole corn tortillas made from instant corn flours by traditional or extrusion processing', *Intl J Food Sci Technol*, 34(4), 391–9.

GOURGUE C, CHAMP M, GUILLON F and DELORT-LAVAL J, (1994), 'Effect of extrusion-cooking on the hypoglycaemic properties of citrus fibre: an in-vitro study', *J Sci Food Agric*, 64(4), 493–9.

GUALBERTO DG, BERGMAN CJ, KAZEMZADEH M and WEBER CW, (1997), 'Effect of extrusion processing on the soluble and insoluble fiber, and phytic acid contents of cereals bran', *Plant Foods Human Nutrition*, 51(3), 187–98.

GUZMAN LB, LEE T-C and CHICHESTER CO, (1992), 'Lipid binding during extrusion cooking', in Kokini JL, Ho C-T and Karwe MV, *Food Extrusion Science and Technology*, New York, Marcel Dekker, Inc.

GUZMAN-TELLO R and CHEFTEL JC, (1987), 'Thiamine destruction during extrusion cooking as an indicator of the intensity of thermal processing', *Intl J Food Sci Technol*, 22(5), 549–62.

GUZMAN-TELLO R and CHEFTEL JC, (1990), 'Colour loss during extrusion cooking of beta-carotene–wheat flour mixes as an indicator of the intensity of thermal and oxidative processing', *Intl J Food Sci Technol*, 25(4), 420–34.

HARPER JM (1981), *Extrusion of Foods*, Boca Raton, FL, CRC Press, Inc.

HARPER JM and JANSEN GR, (1985), 'Production of nutritious precooked foods in developing countries by low-cost extrusion technology', *Food Reviews Intl*, 1(1), 27–97.

HAYAKAWA I (1992), *Food Processing by Ultra High Pressure Twin Screw Extrusion*, Lancaster, PA, Technomic Publ Co.

HAZELL T and JOHNSON IT (1989), 'Influence of food processing on iron availability *in vitro* from extruded maize-based snack food', *J Sci Food Agric*, 46(3), 365–74.

HWANG J-K, KIM C-J and KIM C-T (1998), 'Production of glucooligosaccharides and polydextrose by extrusion reactor', *Starch*, 50(2–3), 104–7.

IBANOGLU S, AINSWORTH P and HAYES GD (1997), '*In vitro* protein digestibility and content of thiamin and riboflavin in extruded tarhana, a traditional Turkish cereal food', *Food Chem*, 58(1/2), 141–4.

INSTITUTE OF MEDICINE (1998), *Dietary Reference Intakes for Thiamin, Riboflavin, Niacin, Vitamin B_6, Folate, Vitamin B_{12}, Pantothenic Acid, Biotin, and Choline*, Washington DC, National Academy Press.

JIN Z, HSIEH F and HUFF HE (1994), 'Extrusion cooking of corn meal with soy fiber, salt, and sugar', *Cereal Chem*, 71(3), 227–34.

KAHLON TS, EDWARDS RH and CHOW FI (1998), 'Effect of extrusion on hypocholesterolemic properties of rice, oat, corn, and wheat bran diets in hamsters', *Cereal Chem*, 75(6), 897–903.

KANG S-Y (1996), *Zinc bioavailability in a semolina/soy protein mixture was not affected by extrusion processing*, M S Thesis, East Lansing, MI, Michigan State University.

KAPANIDIS A N and LEE T-C (1996), 'Novel method for the production of color-compatible ferrous sulfate-fortified simulated rice through extrusion', *J Agric Food Chem*, 44(2), 522–5.

KILLEIT U (1994), 'Vitamin retention in extrusion cooking', *Food Chem*, 49(2), 149–55.

KIVISTÖ B, ANDERSSON H, CEDERBLAD G, SANDBERG A-S and SANDSTRÖM B, (1986), 'Extrusion cooking of a high-fiber cereal product 2. effects on apparent absorption of zinc, iron, calcium, magnesium and phosphorus in humans', *British J Nutr*, 55(2), 255–60.

KLEEMOLA P, PUSKA P, VARTIAINEN E, ROOS E, LUOTO R and EHNHOLM C (1999), 'The effect of breakfast cereal on diet and serum cholesterol: a randomized trial in North Karelia, Finland', *Eur J Clin Nutr*, 53(9), 716–21.

KOKINI JL, HO C-T and KARWE MV (1992), *Food Extrusion Science and Technology*, New York, Marcel Dekker.

KONSTANCE R P, ONWULATA C I, SMITH PW, LU D, TUNICK M H, STRANGE E D and HOLSINGER V H (1998), 'Nutrient-based corn and soy products by twin-screw extrusion', *J Food Sci*, 63(5), 864–8.

LOMBARDI-BOCCIA G, DI LULLO G and CARNOVALE E (1991), '*In-vitro* iron dialysability from legumes: influence of phytate and extrusion cooking', *J Sci Food Agric*, 55(4), 599–605.

LORENZ K and JANSEN G R (1980), 'Nutrient stability of full-fat soy flour and corn-soy blends produced by low-cost extrusion', *Cereal Foods World*, 25(4), 161–2, 171–2.

MAGA J A (1978), 'Cis-trans fatty acid ratios as influenced by product and temperature of extrusion cooking', *Lebensm-Wiss u -Technol*, 11(4), 183–4.

MAGA J A and SIZER C E (1978), 'Ascorbic acid and thiamin retention during extrusion of potato flakes', *Lebensm-Wiss u -Technol*, 11(4), 192–4.

MAHUNGU S M, DIAZ-MERCADO S, LI J, SCHWENK M, SINGLETARY K and FALLER J (1999), 'Stability of isoflavones during extrusion processing of corn/soy mixture', *J Agric Food Chem*, 47(1), 279–84.

MARCHETTI M, TOSSANI N, MARCHETTI S and BAUCE G (1999), 'Stability of crystalline and coated vitamins during manufacture and storage of fish feeds', *Aquaculture Nutr*, 5(2), 115–20.

MARTIN-CABREJAS M A, JAMIE L, KARANJA C, DOWNIE A J, PARKER M L, LOPEZ-ANDREU F J, MAINA G, ESTEBAN R M, SMITH A C and WALDRON K W (1999), 'Modifications to physicochemical and nutritional properties of hard-to-cook beans (*Phaseolus vulgaris* L.) by extrusion cooking', *J Agric Food Chem*, 47(3), 1174–82.

MARTÍNEZ-BUSTOS F, CHANG Y K, BANNWART A C, RODRIGUEZ M E, GUEDES P A and GAIOTTI E R (1998), 'Effects of calcium hydroxide and processing conditions on corn meal extrudates', *Cereal Chem*, 75(6), 796–801.

MARTY C and BERSET C (1988), 'Degradation products of trans-beta-carotene produced during extrusion cooking', *J Food Sci*, 53(6), 1880–6.

MERCIER C, LINKO P and HARPER J M (1989), *Extrusion Cooking*, St. Paul, MN, Am Assoc Cereal Chem.

NELSON A I, WIJERATNE W B, YEH S W, WEI T M and WEI L S (1987), 'Dry extrusion as an aid to mechanical expelling of oil from soybeans', *J Am Oil Chem Soc*, 64(9), 1341–7.

NOGUCHI A and CHEFTEL J-C (1983), 'Extrusion-cooking of protein enriched cookies', *Nippon Shokuhin Kogyo Gakkai Shi*, 30(2), 114–24.

O'CONNOR C (1987), *Extrusion Technology for the Food Industry*, London, Elsevier Applied Sci Publ.

OMUETI O and MORTON ID (1996), 'Development by extrusion of soyabari snack sticks: a nutritionally improved soya-maize product based on the Nigerian snack (kokoro)', *Intl J Food Sci Nutr*, 47(1), 5–13.

OOSTHUIZEN W, SCHOLTZ C S, VORSTER H H, JERLING J C, VERMAAK W J H (2000), 'Extruded dry beans and serum lipoprotein and plasma haemostatic factors in hyperlipidaemic men', *Eur J Clin Nutr*, 54(5), 373–9.

OSMAN M G, SAHAI D and JACKSON D S (2000), 'Oil absorption characteristics of a multigrain extrudate during frying: effect of extrusion temperature and screw speed', *Cereal Chem*, 77(2), 101–4.

PANLASIGUI L N, THOMPSON L U, JULIANO B O, PEREZ C M, JENKINS D J A and YIU S H (1992), 'Extruded rice noodles: starch digestibility and glycemic response of healthy and diabetic subjects with different habitual diets', *Nutr Res*, 12(10), 1195–1204.

POLITZ M L, TIMPA J D, WHITE A R and WASSERMAN B P (1994a), 'Non-aqueous gel permeation chromatography of wheat starch in dimethylacetamide (DMAC) and LiCl: extrusion-induced fragmentation', *Carbohydrate Polymers*, 24(2), 91–99.

POLITZ M L, TIMPA J D and WASSERMAN B P (1994b), 'Quantitative measurement of extrusion-induced starch fragmentation products in maize flour using nonaqueous automatic gel-permeation chromatography', *Cereal Chem*, 71(6), 532–6.

PUTNAM J and GERRIOR S (1999), 'Trends in the U.S. food supply, 1970–1997,' in Frazão E, *America's Eating Habits: Changes and Consequences*,

Washington, Economic Research Service, US Deptartment of Agriculture. Agriculture Information Bulletin No. 750, 133–60.

RALET M-C, THIBAULT J-F and DELLA VALLE G (1991), 'Solubilization of sugar-beet pulp cell wall polysaccharides by extrusion-cooking', *Lebensm-Wiss u-Technol*, 24(2), 107–12.

RIAZ M N (2000), *Extruders in Food Applications*, Lancaster PA, Technomic Publ. Co.

RINALDI V E A, NG P K W and BENNICK M R (2000), 'Effects of extrusion on dietary fiber and isoflavone contents of wheat extrudates enriched with wet okara', *Cereal Chem*, 77(2), 237–40.

SANDBERG A S, ANDERSSON H, KIVISTÖ B and SANDSTRÖM B (1986), 'Extrusion cooking of a high-fibre cereal product. 1. Effects on digestibility and absorption of protein, fat, starch, dietary fibre and phytate in the small intestine', *Br J Nutr*, 55(2), 245–4.

SEMWAL A D, SHARMA G K and ARYA S S (1994), 'Factors influencing lipid autooxidation in dehydrated precooked rice and bengalgram dhal', *J Food Sci Technol*, 31(4), 293–7.

SHIN T S, GODBER J S, MARTIN D E and WELLS J H (1997), 'Hydrolytic stability and changes in E vitamers and oryzanol of extruded rice bran during storage', *J Food Sci*, 62(4), 704–8.

SINGLETARY K, FALLER J, LI J Y and MAHUNGU S (2000), 'Effect of extrusion on isoflavone content and antiproliferative bioactivity of soy/corn mixtures', *J Agric Food Chem*, 48(8), 3566–71.

SOKHEY A S, RIZVI S S H and MULVANEY S J (1996), 'Application of supercritical fluid extrusion to cereal processing', *Cereal Foods World*, 41(1), 29–34.

SRIBURI P and HILL S E (2000), 'Extrusion of cassava starch with either variations in ascorbic acid concentration or pH', *Intl J Food Sci Technol*, 35(2), 141–54.

SUKNARK K (1998), *Development of extrusion processes for novel snack food production and the physicochemical and nutritional properties of resulting products*, PhD Dissertation, Griffin GA, University of Georgia.

SUZUKI H, CHUNG B S, ISOBE S, HAYAWAKA S and WADA S (1988), 'Changes in ω(omega)-3 polyunsaturated fatty acids in the chum salmon muscle during spawning migration and extrusion cooking', *J Food Sci*, 53(6), 1659–61.

THEANDER O and WESTERLUND E (1987), 'Studies on chemical modifications in heat-processed starch and wheat flour', *Starch/Stärke*, 39 (3), 88–93.

TIETYEN J L (1993), *Production and nutritional evaluation of an extruded snack containing soy fiber and plant protein*, Ph.D. Dissertation, Manhattan, KS, Kansas State University.

UMMADI P, CHENOWETH W L and NG P K W (1995a), 'Changes in solubility and distribution of semolina proteins due to extrusion processing,' *Cereal Chem*, 72(6), 564–7.

UMMADI P, CHENOWETH W L and UEBERSAX M A, (1995b), 'The influence of extrusion processing on iron dialyzability, phytates and tannins in legumes', *J Food Process Preserv*, 19(2),119–31.

US DEPARTMENT OF AGRICULTURE, (2000), Table set 17. Food and Nutrient Intakes by Children 1994–96, 1998, 58.

UNLU E and FALLER J F, (1998), 'Formation of resistant starch by a twin-screw extruder', *Cereal Chem*, 75(3), 346–50.

VAN POPPEL G, VERHOEVEN D T, VERHAGEN H and GOLDBOHM R A (1999), 'Brassica vegetables and cancer prevention. Epidemiology and mechanisms', *Adv Exp Med Biol*, 472, 159–68.

WANG W-M and KLOPFENSTEIN C F (1993), 'Effect of twin-screw extrusion on the nutritional quality of wheat, barley, and oats', *Cereal Chem*, 70(6), 712–15.

WANG S, CASULLI J and BOUVIER J M (1993a), 'Effect of dough ingredients on apparent viscosity and properties of extrudates in twin-screw extrusion cooking', *Intl J Food Sci Technol*, 28(5), 465–79.

WANG W-M, KLOPFENSTEIN C F and PONTE J G (1993b), 'Effects of twin-screw extrusion on the physical properties of dietary fiber and other components of whole wheat and wheat bran and on the baking quality of the wheat bran', *Cereal Chem*, 70(6), 707–11.

WENNERMARK B (1993), *Vitamin E retention during processing of cereals*, FILDR Thesis, Lund, Lunds Universitet.

ZADERNOWSKI R, NOWAK-POLAKOWSKA H and RASHED A A (1999), 'The influence of heat treatment on the activity of lipo-and hydrophilic components of oat grain', *J Food Processing Preserv*, 23(3) 177–91.

Part II

Specific extruded products

7

Breakfast cereals

J-M. Bouvier, Clextral, Firminy

7.1 Introduction

Cooked wheat, barley or oats in the form of porridge have been common foods for many hundreds of years, and are still very popular in several countries. However, preprocessed, cereal-based products, like ready-to-eat cereals to be eaten directly out of the packaging, are relatively new. When one speaks of ready-to-eat cereals, or breakfast cereals, the first thought that comes to mind is of cornflakes for breakfast. Since they were first produced more than a century ago by W. K. Kellogg in the United States, cornflakes have initiated a vigorous development of the breakfast cereal industry, and stimulated creativity in marketing and designing breakfast cereal products.

Today, the worldwide consumption of breakfast cereals amounts to about 3 million tons. As shown in Table 7.1, this figure hides large dissimilarities between countries and regions, depending upon their food culture and degree of development. As a matter of fact, the per capita consumption is high in North America and North Western Europe, but the volume growth of the market is rather low (below 5%). The growth mainly concerns breakfast cereals for adults, through healthy diet specialities. In South America and South Western Europe, the per capita consumption is still low (less than 1 kg), while the volume growth ranges from 5 to 20%. Breakfast cereal manufacturers aim to supply products for children and pay attention to taste (sweet flavourings), texture (crispness) and nutrition (vitamins and minerals, in particular). In Eastern Europe, consumption is very low (under 0.10 kg per capita), and breakfast cereals are eaten as bread substitutes; Poland is the largest and the most established market, while Slovakia and the Czech Republic are the fastest growing markets. In Japan and Asia Pacific, the per capita consumption is also very low (below 0.2 kg). It must be

Table 7.1 Worldwide consumption of breakfast cereals and market trends[1–3]

Country, region	Consumption (1997) kg/capita	Yearly market growth Volume (1993–97)	Sales value (1997–2002)	Market trends (1995–98)
United Kingdom	7.6	3 to 4%	< 1%	Growth of healthy and diet specialties for adults (bran-rich products, crispy mueslis, etc.); steady or declining market of breakfast cereals for children.
United States	4.5	3 to 4%	< 1%	
Germany	2.1	4 to 5%	3 to 4%	
France	1.5	6 to 8%	2 to 4%	
Spain	0.75	6 to 8%	< 1%	Growth of consumption by children, mainly due to snacking in particular.
Italy	0.5	15 to 25%	10 to 20%	
Eastern Europe	< 0.10	20 to 25%	–	Bread substitute; exciting new market.
Latin America	0.25 to 1.0	15 to 20%	–	Growth of consumption by children in urban areas; buying habits influenced by health benefits of breakfast cereals (added minerals and vitamins).
Japan	0.18	3 to 4%	3 to 4%	Health market-driven; overall growth limited by the decline of the economy during that period.
Asia Pacific	< 0.10	5 to 15 %	< 0	Low consumption due to cultural issue; market driven by young adults and urban professionals.

noted that the traditional model of ready-to-eat breakfast cereals with milk does not apply, mainly for physiological reasons; lactose is not digested by Asian children. Consumption would only be in the form of snacks, and sweet snacks are not very developed in these areas. Also, the economic problems in Asia during the second part of the nineties depressed consumer markets in general, and hence the consumption of breakfast cereals. However, some countries in this region like South Korea, Taïwan and Singapore, do offer good opportunities for breakfast cereal manufacturers, and the per capita consumption will probably increase in the coming years, in urban areas in particular.

The popularity of breakfast cereals really stems from their nutritional content. Such products can simultaneously provide energy (350–400 kcal/100 g), nutrients (vitamins, minerals) and health-oriented components (dietary fibre, for instance). Throughout the long history of breakfast cereals, advertising has always underlined the health potential of the products, and consumers do generally recognise such beneficial input. Today, the impact of nutrition on health and disease is becoming a major concern of food designers, food manufacturers and consumers. Thus, the healthy eating trend increasingly influences food choice, and breakfast cereals do not escape this development. Of course, the health issue leads to innovation when designing breakfast cereals.

Over the last three decades, extrusion-cooking technology has played a very important and decisive role in the innovation and development of breakfast cereal products. The extrusion-cooking process is a new cooking concept – thermomechanical HTST (high temperature short time) cooking – that makes particular use of mechanical processing of the material and is therefore an original alternative to classical hydrothermal cooking. Then, it allows continuous cooking of a large range of recipes with various cereals to produce different shapes and textures at satisfactory cost. Two types of extrusion-cooked breakfast cereals can thus be found on the market:

- *Directly expanded extrusion-cooked* breakfast cereals. Cereal flours and/or grits are cooked with ingredients and with a very low moisture content (usually below 20%). The process may use single- or twin-screw extruders, the configuration and operating characteristics of which generally lead to highly mechanical cooking.
- *Pellet-to-flakes extrusion-cooked* breakfast cereals. Cereal flours and/or grits are cooked with ingredients and at a moisture level in the range of 22–26%. They are usually processed in twin-screw extruders, the configuration and operating characteristics of which lead to a lower mechanical component of cooking, reinforcing the thermal component as opposed to the previous processing conditions.

This chapter begins with an overview of market segments, together with breakfast cereal products which are on the market. Then, it focuses on those products made by extrusion-cooking technology, describing and discussing extensively the two generic extrusion-based processes. An important part of the chapter is dedicated to describing and discussing the major unit operations and technologies of both processes: engineering analysis of process functions, equipment design issues and their effect on product quality.

7.2 The range of products

The breakfast cereal section of supermarket shelves displays an impressive range of products, which differ in their packaging design and sizes, as well as product characteristics (nutritional value and sensory attributes, in particular). The major manufacturers are competing to satisfy three important segments of the population:

- children
- nutrition- and health-conscious consumers
- fitness-oriented consumers.

The market for children needs breakfast cereals which offer a large diversity of tastes (honey, chocolate and malt), shapes (balls, cups, animals, puffed grains, flakes, etc.), textures (from dense-hard to crispy-soft) and colours. These products are generally packaged in colourful boxes displaying fancy animals and characters. These requirements need to be balanced with those of the purchasing parent for a healthy and nutritious product.

Adults show much more interest in functional breakfast cereals which contribute to health preservation and fitness build-up. Typical products consist of traditional and crispy mueslis, mixes of cereals with nuts and fruits, bran-rich products, and so on. Breakfast cereals are natural sources of complex carbohydrates, fibre and the B-group of water-soluble vitamins. Also, modern manufacturing techniques like extrusion-based processes, make it possible to complement the natural components of breakfast cereals with other groups of vitamins (A, C, D, E), minerals (iron, calcium, phosphorous, magnesium, zinc) and fibres (brans, soluble dietary fibre), to meet the expectations of the adult segment of the market.

Market-driven factors and breakfast food cultures have both contributed to promote three generic ready-to-eat breakfast cereals which are found on supermarket shelves nowadays (Table 7.2):

- flaked cereals
- puffed cereals
- cereal mixes.

In view of their historical character, flaked cereals must be mentioned first. In fact, traditional cornflakes have good specific characteristics (crispy texture and bubbled flat shape), and consequently a strong position on the market. Flaked cereals (mainly corn and wheat flakes) traditionally use *flaking grits*, that are raw pieces of grain endosperm. Flaking grits are hydrothermally cooked in a batch steam cooker, then flaked and toasted to obtain the well-known texture and shape. Flaked cereals can also be manufactured by extrusion-cooking.

Table 7.2 Generic breakfast cereals. Process characteristics and product specificities

Breakfast cereals	Main process characteristics		Main product specificities				
Generic products	Raw materials	Cooking	Texture	Shape	Taste	Nutrition and Health	Fitness
Flaked cereals	Flaking grits (corn, wheat)	Hydrothermal (batch steam cooking)	X	X			
	Flours (corn, wheat, oat)	Thermomechanical (continuous extrusion-cooking)	X	X		X	
	Flours and bran (corn, wheat, oat)	Thermomechanical (continuous extrusion-cooking)	X	X		X	
Puffed cereals	Whole grains (wheat, rice)	Hydrothermal (batch steam cooking)	X				
	Flours, grits and bran (any cereals)	Thermomechanical (continuous extrusion-cooking)	X	X	X	X	
Cereal mixes	Ground grains, nuts and dried fruits	–			X	X	X
	Ground grains, nuts and dried fruits	Thermal toasting	X		X	X	X

Table 7.3 The major directly expanded, extrusion-cooked breakfast cereals on the market

Product	Shape	Cereals	Texture	Flavour
Ball		corn	crispy, highly expanded	honey, chocolate
Rice shape		rice	crispy, low hardness	neutral, chocolate
Loop		wheat, oat, corn	crispy, relatively hard	neutral, honey
Cup		wheat	hard, dense structure	chocolate
Cylinders	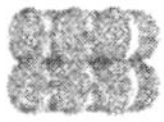	corn	crispy	honey
Bran stick		wheat and bran	crunchy and hard, dense structure	malt
Heart shape		wheat and oat bran	crunchy and hard, dense structure	malt and honey

Rather than specific flaking grits, any cereal flours including bran can be thermomechanically cooked in a complete continuous process. Extrusion-cooking technology then makes it possible to use cheaper raw materials as well as to increase processing line productivity, and finally to decrease the price of flaked cereals. Though the quality profile of the flakes produced from traditional and extrusion-based processes is significantly different (texture in particular), owing to the different cooking characteristics of the processes, extrusion-cooking technology has definitely helped promote basic flakes (corn and wheat flakes) as well as to develop speciality flakes like nutrition- and health-oriented flakes (bran flakes and oat-rich flakes, for example).

Traditional puffed cereals are produced by gun puffing precooked whole grains (hard wheat or durum, long-grain white rice or parboiled medium grain rice). Gun puffing is a batch steam-induced expansion process, leading to crispy products the shape and texture of which are defined by the type of raw material. The process is in fact quite limited since texture and shape of final products no longer vary. A direct expansion extrusion-cooking process makes it possible to use various raw materials and recipes, and thus to modify product characteristics extensively. The process is continuous and flexible, and offers real ways of optimising product quality and process productivity. Table 7.3 shows the major directly expanded, extrusion-cooked breakfast cereals which are found on the market today. Such products provide energy (370–400 kcal/100 g) and have different textures, shapes, colours and flavours; they are particularly adapted to the 'children' segment of the market. The direct expansion extrusion-cooking process can also texturize nutritious bran-rich and oat-based breakfast cereals.

Traditional mueslis generally consist of a mixture of various natural components: ground whole cereal grains (wheat, barley or oats), oilseeds (sesame and sunflower seeds), nuts (almonds, coconut, walnut), dried fruits (raisins, apple, banana, etc.). Such mixes are a good compromise between the energy content (around 350 kcal/100 g) and the nutritional potential (10 to 12% proteins, 8 to 10% fibres, relatively low sugar content under 15%, balanced lipid composition) of the food. Crunchy and crispy mueslis are having more and more success with adult consumers, as they offer a wide diversity of products, from high energy to healthy mixes. Such products are usually composed of toasted ground cereal grains, together with nuts and dried fruits. Toasted ground cereal grains may also be sugar coated and flavoured (chocolate, honey, malt, cinnamon).

In addition to traditional ready-to-eat breakfast cereals (flakes and cereal muesli mixes) which offer low product diversity and flexibility, an impressive range of new ready-to-eat breakfast cereal products have been developed over the last two to three decades to respond to the consumer's demand for convenience, pleasure, nutrition and health. Advanced technology, like extrusion-cooking, has made a considerable contribution to satisfying this demand by offering continuous and flexible processes, allowing the breakfast cereal manufacturers to process various recipes and develop a wide range of products with diverse textures and shapes, and ultimately to reduce the cost of final products.

7.3 Key process issues of the product range

The cooking characteristics of extrusion technology enable breakfast cereal manufacturers to process any cereals and starch-based recipes, at high productivity and product diversity. Compared with conventional, batch hydrothermal cooking at 90–110°C with 15–30 minutes residence time, continuous thermomechanical cooking as applied in extruders operates at higher temperatures (140–180°C) with a shorter residence time (0.5–1.5 minutes). This is crucial in making extrusion-cooking processes much more productive, due to mechanical shear in particular. Also, although hydrothermal and thermomechanical cooking processes both convert starch polymers from a semi-crystalline to amorphous state, there are still substantial differences between the two types of cooking. The plasticizing effect of extrusion leads to specific macromolecular structures of starch polymers. Moreover, as the extent of starch dextrinization is negligible in hydrothermal cooking, it is very sensitive to the shearing effect present in extruders, which helps vary the distribution of the molecular weights of starch macromolecules. Thus, by combining variations in the plasticizing effect and the dextrinization extent in extrusion, it is possible to obtain various polymer structures and molecular weight distributions, leading to a broad diversity in the functional properties of extruded polymeric melts. Finally, extrusion-cooking technology offers wide potential for varying the functional properties of polymeric melts, and hence the characteristics of final breakfast cereal products.

As already mentioned, two types of extrusion-cooked breakfast cereals are found on the market, made with two generic extrusion-cooking processes nowadays used by breakfast cereal manufacturers:

- direct expansion extrusion-cooking process (the so-called DEEC process)
- pellet-to-flaking extrusion-cooking process (the so-called PFEC process).

7.3.1 Direct expansion extrusion-cooking

In this process, the extruder not only cooks the raw materials, but also texturizes and shapes the final products. Figure 7.1 shows a typical flowsheet of the DEEC process, which mainly consists of five successive unit operations:

- mixing of raw materials and base ingredients
- extrusion-cooking
- drying – which may include toasting
- syrup coating
- drying/cooling.

The dry raw materials, usually a mixture of flours (corn, wheat, rice or oats) and ingredients (modified starches, bran, sugar, emulsifiers, sodium chloride, calcium phosphate, etc.), are gravimetrically metered in a batch blender and then premixed. A horizontal, batch blender whose agitator has two reverse-spiral ribbons giving uniform mixing, is commonly used. The premixed recipe is fed into a circular buffer bin and transferred to the extruder feeder by a screw conveyor.

The feeder delivers a uniform, continuous flow of premixed recipe into the extruder. Two types of feeder are normally used: a volumetric screw-type feeder and a loss-in-weight feeder. The more common and cheaper type is the volumetric feeder where the feed is proportional to the speed of screw rotation and thus depends upon the bulk density of the premix. The feeder has either a single screw or self-cleaning twin screws; the twin-screw feeder delivers more constant flow rate than the single-screw feeder, and it handles fairly sticky materials better with greater accuracy. If more precision is required in the feed flow, a loss-in-weight feeder is used to provide a constant mass flow rate of dry premix to the extruder.

The extruder cooker in the processing line may be a single-screw or an intermeshing co-rotating twin-screw extruder. The mixture is processed in a short extruder with a relatively simple screw configuration which has a cooking section in the terminal position. In this section, the material is intensively sheared in the molten state. The water needed for the cooking process comes from the raw materials and moisture adjustments are made by a volumetric pump directly into the feed section of the extruder; the total moisture content in the extruder then ranges from 16 to 20%. The screw speed is usually set between 200 and 450 rpm which, combined with the screw profile and moisture content, makes it possible to adjust shear stresses and mechanical work in the cooking section. These processing conditions mean that the extruder cooks the material

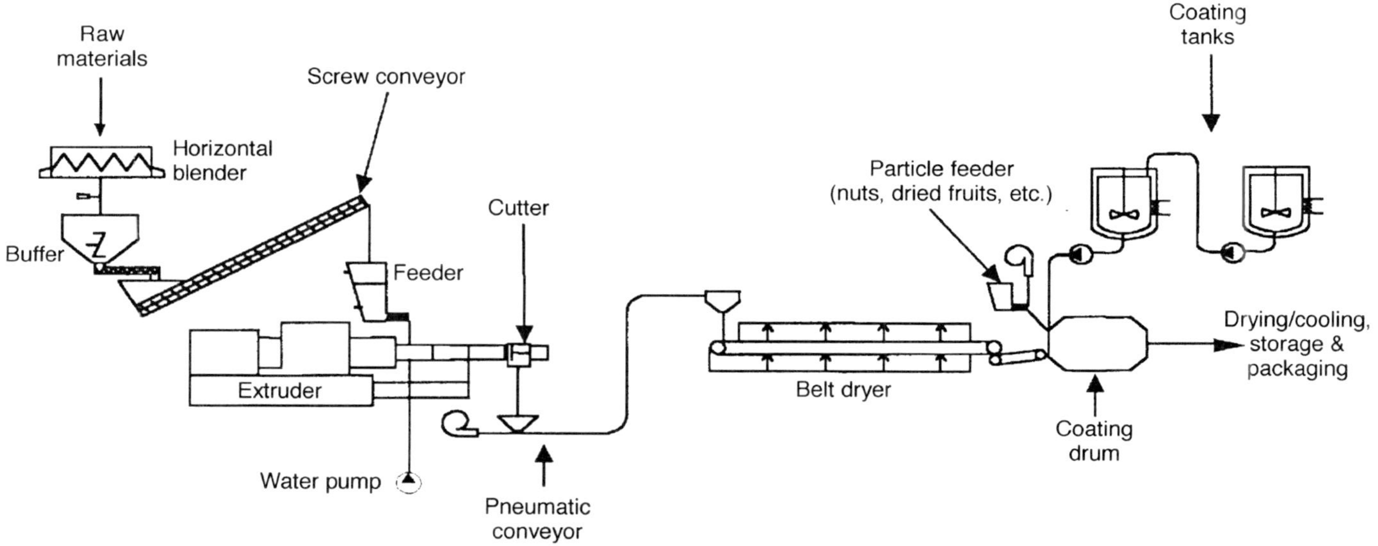

Fig. 7.1 Typical flowsheet of direct expansion extrusion-cooking process.

more mechanically than thermally. Malt syrup may be added directly to the melt just as it enters the cooking section.

The extruder also has a die assembly at the end of the screw-barrel system, through which the product is extruded. The cooked melt in the die at high pressure (above the vapour pressure of water) and high temperature (above 100ºC) undergoes considerable expansion on exiting the die, owing to the flash pressure reduction and the evaporation of the water present: the melt expands directly at the die. Water evaporation causes a rapid fall in the temperature of the melt, which becomes more and more viscous. It hardens quickly into a highly aerated structure that gives the final product a very pronounced crispy/crunchy texture. The rate of expansion depends on the rheological and thermal properties of the molten material and on the geometry of the shaping insert. The action of the die is supplemented by a cutter which directly cuts the expanding strands as they leave the die, so as to give the desired shape to final products. The die assembly is a very important piece of the extruder as it determines the quality profile of expanded breakfast cereals and product consistency (bulk density, texture and shape, in particular). Table 7.4 summarizes the main processing conditions (extrusion-cooking and die texturization) of major directly expanded extrusion-cooked breakfast cereals.

The extruded products are transferred to the drier after cutting either by a belt elevator or through a pneumatic conveyor. At the drier inlet, the extruded products have a moisture content of 7 to 10%; this must be reduced to 2 to 3% to give the right crispness as they leave the drier. Drying takes place continuously on single-pass or multipass conveyor belt driers, with heat transfer by convection. The drier may be fitted either with electrical resistances, a gas burner, or even steam, to heat the ventilating air. The drying temperature is usually between 140 and 160ºC; the residence time is optimized by varying the belt speed, and ranges between 3 and 8 minutes. The orientation of the product changes from pass to pass, and hence its exposure to the drying air, which results in more uniform drying. The drier may have a toasting section for some browning of breakfast cereals (crisp rice, for example); this occurs at temperatures ranging between 150 and 200ºC. Fluidized-bed driers may also be used to dry and toast expanded breakfast cereals.

Dried expanded breakfast cereals are often coated with flavoured sugar syrup (about 80ºBrix). The products are then blended in a cylindrical rotating drum, and exposed to the spray of coating syrup. The drum is designed to create a folding action in the product bed, to facilitate contact between the product and syrup and ensure a uniform coating. The flavoured sugar syrup is handled in two jacketed tanks in series: syrups are batch prepared in the first tank, while the second tank continuously feeds the spray system of the drum. Finally, the sugar coated breakfast cereals are dried and cooled in a conveyor belt drier before packaging.

7.3.2 Delayed expansion extrusion-cooking

This process is used to manufacture pellet-to-flakes extrusion-cooked breakfast cereals. In this case, the extruder only cooks the raw materials, producing pellets

Table 7.4 The main processing conditions of major directly expanded, extrusion-cooked breakfast cereals

Product		Extrusion-cooking		Die texturization	
	Screw speed rpm	Barrel temperature °C	SME kJ/kg	Insert mass flux $kg/mm^2.h$	Bulk density g/l
Ball (corn-based)	300–450	130–150	400–450	4–5	40–60
Crisp rice (rice-based)	300–400	160–180	380–450	3–4	110–120
Loop (oat-based)	200–300	140–160	320–400	5–6	180–220
Cup (wheat-based)	250–350	110–130	620–700	3–4	120–140
Bran stick	200–300	115–135	550–620	2–3	180–200

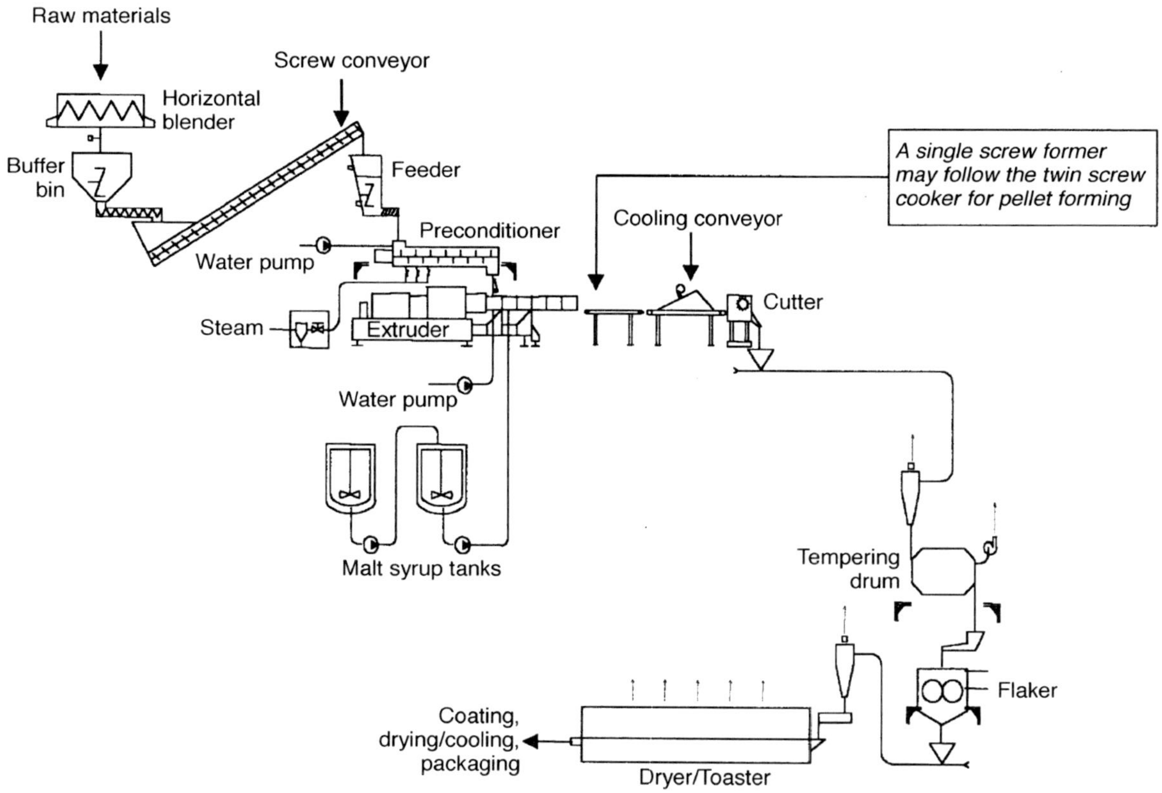

Fig. 7.2 Typical flowsheet of pellet-to-flaking extrusion-cooking process.

for flaking. Figure 7.2 shows a typical flowsheet of the PFEC process, which mainly consists of seven successive unit operations:

- mixing of raw materials and base ingredients
- thermomechanical cooking in the extruder
- pellet forming
- pellet flaking
- drying/toasting
- syrup coating
- drying/cooling.

The mixing of the dry raw materials (corn flour, corn grits, wheat flour and oat flour) and the base ingredients (bran, sugar, emulsifiers, etc.) is identical to that of Fig. 7.1 (see section 7.3.1).

The mix enters the cooking section of the process, through either a volumetric or a loss-in-weight feeder. The cooking section consists first of a preconditioner which heats and humidifies the mix. The preconditioning process takes place continuously either in a single shaft or in a counter-rotating twin shaft system. The shafts are fitted with paddles the design and configuration of which, together with shaft rotation speed, make it possible to adjust operating conditions: filling ratio, residence time distribution and efficiency of mixing. The mixture is humidified by spraying water downwards into the preconditioner through spray nozzles. The mixture is preheated by steam injection through a row of injectors directed upwards. The mix then reaches a temperature of 75–85°C and a moisture content of 18–20%. Preconditioning is not essential in the process, but thermomechanical cooking would then be significantly different when preconditioning the mix.

The preconditioner feeds the extruder cooker which is usually an intermeshing co-rotating twin-screw extruder to handle operating conditions with a high level of flexibility and able to have a better consistency of cooking, as opposed to a single-screw extruder. The mixture (starch-based materials and various ingredients) is processed in a long extruder with an L/D ratio of between 20 and 30. It has a relatively complex screw configuration with several cooking sections in series, each composed of screw elements giving moderate shear; low pitch conveying screw elements are placed between cooking sections. The screw configuration ends with a transport section that is designed to facilitate heat transfer between the material and the barrel, in order to cool the molten material (below 100°C) before shaping in the die. This last section can include a degassing barrel, which is designed particularly to increase the cooling efficiency. The screw speed rarely exceeds 200 rpm; combined with a relatively high water content (from 22 to 26%) and a screw configuration giving reduced shear, this has the overall effect of satisfactorily handling the mechanical and thermal components of cooking. Malt syrup is usually added directly to the extruder through the feeding section, or the early stages of the cooking section. The molten material is then fed into a multi-hole die assembly in order to shape and cut vitreous, unexpanded or very slightly expanded pellets (5–6 mm diameter, 6–8 mm long). These can be produced either by die-face cutting, or delayed cutting where the die produces ropes which are then cut. After cutting, the temperature and moisture content of the pellets are 80–95°C and 20—22% respectively. Next, pellets are transferred to a tempering drum for cooling (40–60°C), and also to prevent the sticking of pellets and to regularize the internal moisture gradient. The extent of cooling is controlled by a counter-current flow of ambient air in direct contact with the pellets. The drum rotates at a constant speed while it is slightly tilted, so as to give a plug flow behaviour and control the dwell time before flaking.

The L/D ratio of the twin-screw extruder cooker may be limited and the extruder used just as a cooker. In this case, the hot cooked melt produced by the twin-screw extruder is fed into a slowly turning single-screw former. This gently kneads and cools the melt, and allows its temperature to be controlled for die-face cutting. This optional process configuration brings more flexibility as it decouples cooking and pellet forming.

On leaving the drum, the pellets are transferred to a vibratory feeder which supplies single pellets in a uniform and continuous flow, and disperses the pellets across the whole width of the flaker. The pellets then fall directly into the rolling zone where they are flattened. The flaker consists of two rollers of the same size; one of them being movable so that the space between the rollers can be adjusted, while keeping them parallel, in order to regulate the thickness of the flakes. When the flaker is operating, the rollers tend to heat up; care must therefore be taken to ensure that they are constantly cooled by an internal cooling system, to keep the space between the rollers constant. Scraper blades separate the rubbery flakes from the roll surface.

Rubbery flakes enter the drying-toasting unit at a temperature and moisture content of about 32–38°C and 18–20% moisture content, respectively. Because of the uniform heat transfer and flake treatment at relatively short dwell times, air-impingement technology is much preferred to traditional rotary or conventional conveyor belt technologies. An air-impingement drier has generally three sub-units: the first one uses high temperatures (220–270°C) to remove moisture and puff the flakes, while the second one acts as a toaster at temperatures ranging from 160 to 200°C, so as to give the specified expansion, crunchiness and colour to the flakes. The third sub-unit cools the puffed, toasted flakes to stop the toasting process, and prepare the product for further processing or packaging. Additional processing may include coating with flavoured sugar syrup in a cylindrical rotating drum, and then the coated flakes are dried and cooled, as described in section 7.3.1.

7.4 Main unit operations and technologies

The two generic extrusion-cooking processes presented above both involve important unit operations, the engineering characteristics and relative technologies of which play a fundamental role in product quality (surface aspect and shape and texture, in particular). Those quality factors depend strongly on thermomechanical cooking including preconditioning, die texturization, and drying including toasting if required. It is therefore important to analyse those unit operations in depth and particularly the physical mechanisms which govern their process functions and the related design of selected technologies:

- preconditioning
- extrusion-cooking
- drying and toasting.

7.4.1 Preconditioning

Predicting the hydration time of starch-based flours and grits

Preconditioners are used to preheat and prehumidify biopolymer-based raw materials such as flours and grits by mixing them with steam and water. During preconditioning therefore, heat and water must be uniformly distributed within particles to avoid temperature and moisture gradients before feeding and cooking in the extruder. This prompts a basic investigation of the heat and mass transfer between components of the three-phase gas (steam)/liquid (water)/solid (flours, grits) medium inside the preconditioner.

In practice, when the slightly superheated steam and liquid water are fed into the preconditioner, steam helps to heat the particles due to its condensation and, together with water, generates a thin film of water around the flour and grit particles. Thus when the cold particles are surrounded by the hot saturated steam, the temperature and moisture content of the particles increase, although not instantaneously. Two factors govern the rate of heating and swelling of the particles. The first is the film resistance at the surface of the particle; this relates to the quality of the contact between the fluid and the particles. The better the fluid/solid contact, the lower the film resistance. The mixing efficiency of the preconditioner will then determine the film resistance. The second factor is the rate of heat and moisture flows into the interior of the solid particle. This is the internal resistance governed by Fourier's second law and Fick's second law, respectively. Knowing the diffusivity coefficients, these physical laws can be used easily to predict the time necessary to heat and humidify the particles homogeneously. In general, the higher the heat and water diffusivity, the higher the rate of heat and moisture flows in the particles.

In the case of starch materials at ambient temperature, the thermal diffusivity is about $10^{-7}\,m^2/s$; while the water diffusivity is nearly $10^{-9}\,m^2/s$. Thus, the thermal diffusivity is 100 times larger than the water diffusivity, meaning that heat transfer is much faster than moisture transfer. Accordingly, moisture transfer controls the process of preconditioning. Solutions of Fick's second law can be expressed through the non-dimensional Fourier number, Fo, as follows:

$$Fo = \frac{Dt}{(R/3)^2} \tag{7.1}$$

where D, t and R are the water diffusivity, the diffusion time and radius of the particle, respectively.

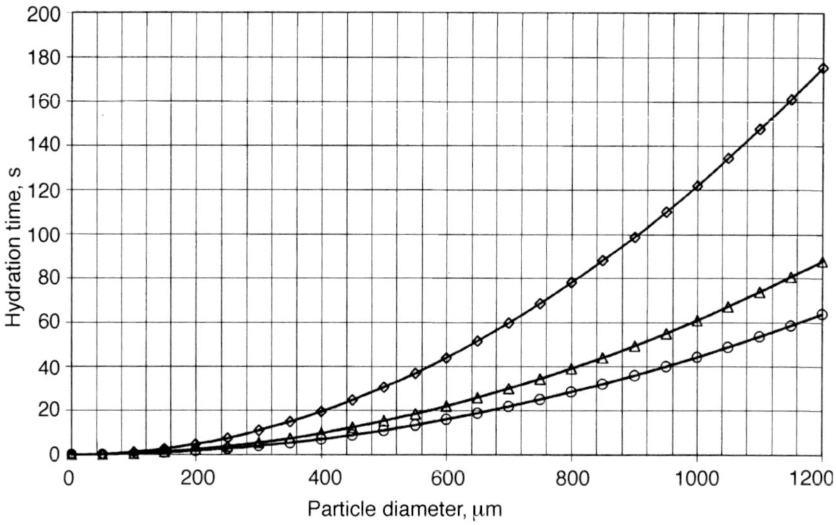

Fig. 7.3 Hydration time as a function of particle size (◇: 60°C, △: 80°C, ○: 90°C; Co = 0.99 Cs; Bi = 1; from Bouvier[8]).

The relative importance of the film and internal resistance terms is measured by the dimensionless Biot number (*Bi*). For a small Biot number ($Bi < 0.1$), the main resistance is in the film around the particle; this would be the case in preconditioners with poor mixing. For a high Biot number ($Bi > 10$), the main resistance is diffusion of water into the particle; this would be the case in preconditioners with good mixing. Preconditioners currently used in industrial processes would show intermediate efficiency of mixing, meaning that *Bi* is close to 1; both film and internal resistance terms affect the moisture transfer in the particle.

Solutions of the mass balance equations that account for both film and internal resistances have been derived and presented in various engineering books related to heat and mass transfer.[4, 5] Such solutions allow the evaluation of the Fourier number as a function of Biot number and moisture concentration at the centre of the particle (*Co*) as opposed to the moisture concentration at the surface (*Cs*). Allowing for the water diffusivity in starches at 60°C, 80°C and 90°C,[6, 7] and assuming that $Co = 0.99\,Cs$ (negligible moisture gradient in the particle), it is possible to calculate the diffusion time *t* (or hydration time: the time for homogeneous hydration of the particle) as a function of particle diameter by using equation 7.1. Figure 7.3 shows the calculated data that represent the minimum time required to prehumidify the particles at various temperatures (60°C, 80°C and 90°C), when film and internal resistance terms affect the moisture transfer in the particle which would represent normal mixing efficiency in preconditioners ($Bi = 1$).

Residence time distribution in preconditioners

In reality, preconditioners offer various processing times and degrees of mixing according to their operating conditions (for example shaft speed and throughput) and paddle configuration. This can be investigated by determining the Residence Time Distribution (RTD). In fact, the determination of the RTD provides two important answers. The first is the time history offered by the preconditioner, which is used to obtain the required hydration time. The second is the degree of axial mixing, which contributes to the effectiveness of the fluid/particle contact in the preconditioner. The extent of radial mixing also affects this contact; it depends on shaft rotation speed and paddle geometry.

Although preconditioners have long been associated with extrusion-cooking processes, engineering studies like RTD analysis in preconditioning are still very rare. Recently, Bouvier investigated the RTD in a twin shaft counter-rotating preconditioner.[8] This study showed that flow behaviour in the preconditioner ranged between plug flow and perfectly mixed flow (high axial mixing), depending on operating conditions (shaft rotation speed and paddle configuration, in particular). For example, increasing shaft rotation speed from 60 to 200 rpm, shifted the flow behaviour from ideal plug flow to perfectly mixed flow.

In preconditioning, the RTD can be represented by the classical tanks-in-series model which involves two important parameters:[8] ts, the average residence time and J, the number of perfectly mixed tanks in series. Actually, the value of J reflects the degree of axial mixing in the preconditioner. The highest degree of axial mixing theoretically corresponds to $J = 1$, where the distribution shows very large dispersion of residence time; this is called perfectly mixed flow. In contrast, when $J \rightarrow \infty$, the flow behaves as a plug flow with no axial mixing and no dispersion of residence time.

The practical implication of RTD is preconditioning efficiency. If the distribution is broad, some of the particles may spend some short time in the preconditioner and receive insufficient moisture. This is a particular problem when the raw materials show a wide dispersion of particle sizes. In fact, shorter times may generate important moisture gradients within the particles, particularly the biggest ones, resulting in particles with various moisture contents; this would introduce further cooking heterogeneities in the extruder. Intermediate dispersions of RTD must be applied in preconditioning, to avoid too short times and to obtain sufficient axial mixing. This means values of 5–7 for J. It must be noted that the efficiency of mixing also depends upon radial mixing which is particularly affected by shaft rotation speed and paddle design, which will be discussed in the next section.

Preconditioners and preconditioning conditions

Breakfast cereal manufacturers mostly use twin shaft counter-rotating preconditioners, which offer much better mixing and hence more homogeneous prehumidification compared with single shaft preconditioners. Preconditioners are characterized primarily by their free volume, filling ratio and specific capacity. The filling ratio expresses the ratio of the mix volume (based on mix bulk density) to the free volume; the mix volume being measured when the preconditioner is stopped. The specific capacity expresses the ratio of the mass flow rate of mix to the free volume. In the pellet-to-flakes extrusion-cooking process, the filling ratio of the preconditioner normally ranges from 15 to 35%, giving a specific capacity of 1.5 to 4 kg/k.L.; the average residence time then ranges from 2 to 3 minutes.

The preconditioning conditions are governed by paddle configuration. For instance, twist paddles make it possible to increase the axial mixing, while flat paddles can increase the filling ratio (up to 50–60%, if required). Paddles are generally screwed on shafts, and this enables operators to adjust paddle angles with great flexibility. Paddle angle results from the orientation of the paddle width with respect to the shaft axis direction: an angle of 0º is obtained when paddle width is parallel to shaft axis; this position is called neutral which offers very good radial mixing, but no axial mixing. Varying paddle angle in any direction then makes it possible to vary the intensity of the axial flow of particles in the preconditioner (forward flow and reverse flow, when paddle angle is positive and negative, respectively); this makes it possible to combine axial and radial components of mixing. Paddle configuration results from the arrangement of forward and reverse sections finally to optimize the time history, the filling ratio and the mixing efficiency in the preconditioner. In the pellet-to-flakes extrusion-cooking process, twin shaft preconditioners equipped with twist paddles are preferred when alternating forward, neutral and reverse flow sections; in forward and reverse flow sections, paddle angle normally varies between ±20º and ±45º.

7.4.2 Extrusion-cooking

Thermomechanical cooking of breakfast cereal recipes

The extrusion-cooking unit thermomechanically cooks breakfast cereal mixes to generate the right functional properties of the resulting melt according to the quality profile of the final products. This involves shear- and thermal-induced conversion of biopolymers, at relatively high temperature (140–180ºC) and short dwell time (20–60 s), which leads to particular properties of the cooked material compared with conventional hydrothermal cooking. By varying the relative contribution of the mechanical and thermal components of extrusion-cooking, it is possible to process numerous recipes and products, and to reproduce the characteristics required by ready-to-eat breakfast cereals. However, the extruder-cooker must be properly designed to obtain the required extent of thermomechanical cooking.

The extruder-cooker is characterized primarily by its L/D ratio (barrel length to screw diameter). In the direct expanded extrusion-cooking process (DEEC process), the L/D ratio ranges between 9 and 15; while in the pellet-to-flaking extrusion-cooking process (PFEC process), it ranges between 20 and 30. In both processes, the screw configuration is normally composed of feed and compression sections, where the material is heated by interparticular friction and conductive heat transfer until melting occurs; the material changes from a solid particulate state to a continuum (viscous fluid), and the compression section can advantageously terminate in a short mixing section (neutral 90° position, or forward 45° position) to complete biopolymer melting. The molten material then enters the cooking section which is fitted with screw elements of high shear profile in the DEEC process (very low channel depth in the single-screw extruder and reverse pitch in the twin-screw extruder), and with screw elements of low shear profile in the PFEC process (mixing discs in neutral 90° position in twin-screw extruder). In the cooking section, part of the mechanical energy is dissipated and converted into heat, while the rest is used to convey and convert the biopolymeric material mechanically.

Breakfast cereal extrusion-cooking processes involve low moisture contents (below 25–26%) and high temperatures (above 130–140°C), while substantial shear forces are applied. In such conditions, starch granules undergo not only gelatinization but also melting. Owing to the substantial contribution of shear forces, starch is thereby converted in much shorter times than would be possible using heat alone. Starch conversion involves particularly the loss of granule integrity, loss of ordered regions in each granule, a reduction in biopolymer molecular weight, and the formation of amylose-lipid complexes. These changes together might represent thermomechanical cooking, the extent of which is determined by the well-known process response, called the Specific Mechanical Energy (SME). The SME is the ratio of the net mechanical energy input W (W can be derived from the drive power), to the total mass flow rate Q; if the mass flow rate is expressed in kg/h, $SME = 3.6\,W/Q$ (in kJ/kg).

The SME is a key process parameter. In fact, it is used to correlate extrusion-cooking conditions (screw speed, moisture content, screw configuration, etc.) and product conversion (Water Absorption Index, WAI; Water Solubility Index, WSI), and thus to optimize and scale up the process appropriately. Smith[9] has investigated microstructural changes in starch materials as a function of screw configuration, moisture content and barrel temperature, achieving a range of SME values from 180 to 750 kJ/kg (experiments carried out with maize grits). Smith's data demonstrate a general relationship between WAI and WSI, as shown in Fig. 7.4. In fact, the quantity of swollen starch granules increased with increasing SME; but starch granules were undamaged when the SME remained below 350–400 kJ/kg. Also, WAI increased as SME increased from 180 to 350–400 kJ/kg, due to an increasing proportion of gelatinized starch granules; starch solubility also increased with increasing SME, because of macromolecular degradation of starch. The WAI–WSI relationship reached a maximum at an intermediate level (350–400 kJ/kg $<$ SME $<$ 500–550 kJ/kg), where damaged and undamaged starch granules were found simultaneously in the melts: appearance and disappearance of gelatinized starch are then balanced. As SME increased from 500–550 kJ/kg to 750 kJ/kg, starch granules were totally damaged; WAI decreased and WSI increased. Though such a relationship is general for starchy materials, the absolute values of WAI, WSI and SME may of course change, when processing real recipes or different starch types. The domains of both the DEEC and PFEC processes can then be located in Fig. 7.4: the PFEC process would tend to be to the left of the WAI peak (SME $<$ 400 kJ/

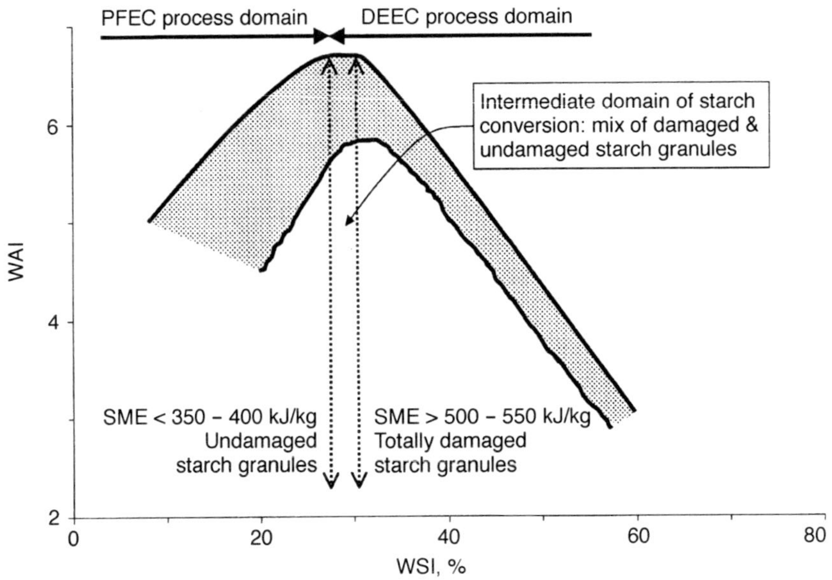

Fig. 7.4 WAI–WSI relationship in extrusion-cooking of maize grits (from Smith[9]).

kg); while the DEEC process would occur at the maximum of WAI and beyond (SME > 400 kJ/kg). In each domain, changing the extrusion-cooking conditions (screw speed, screw design, moisture content and barrel temperature) makes it possible to vary the thermomechanical cooking extensively (SME) and hence, the physico-chemical characteristics of starch (WAI, WSI).

The use of the RVA method (Rapid ViscoAnalyser) in this field must be mentioned, as it may improve the analysis of thermomechanical cooking in extrusion-cooking processes. Figure 7.5 shows RVA profiles of corn-based extrudates produced from a DEEC-type process carried out in a Clextral co-rotating twin-screw extruder; L/D: 16; barrel diameter: 25 mm. Two different reverse screw elements (25 mm long) were used in the cooking section of the screw profile: screw profile 1 with 4 half-moon slots, conjugated reverse screw element (reverse screw diameter: 24.75 mm), and screw profile 2 with free-slot conjugated reverse screw element (reverse screw diameter: 23.6 mm); the reverse screw elements are designed so as to offer exactly the same equivalent cross section when the melt flows through. Also, two different moisture contents (14 and 17%) and screw speeds (250 and 400 rpm) were used. RVA patterns of extrudates typically show large differences when compared to that of raw material: a considerable increase in cold viscosity (left peak between 0 and 6 minutes) with concomitant disappearance of raw peak at about 10.5 minutes, and a decreased setback viscosity at 25 minutes. The disappearance of the raw peak means that starch in the extrudates is totally gelatinized; while the decrease of setback viscosity of the extrudates is due to starch degradation. As shown in Fig. 7.6, the cold viscosity peak is well correlated with SME: it is inversely

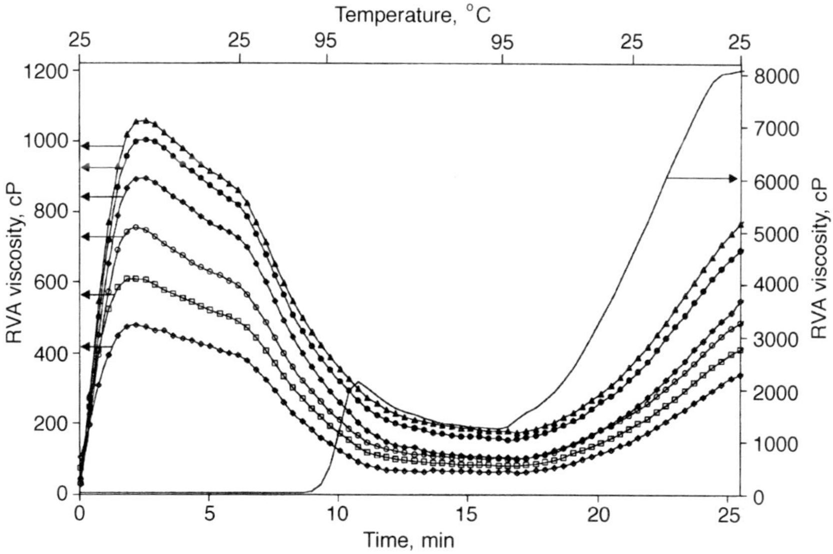

●: LS-HW-SP1; ◆: LS-LW-SP1; ◇: HS-LW-SP1; ○: HS-HW-SP1; ❑: HS-LW-SP2; ▲: LS-HW-SP2; —: corn, raw material).

(LS: low screw speed, 250 rpm; HS: high screw speed, 400 rpm; LW: low water content, 14%; HW: high water content: 17%; SP1: screw profile 1; SP2: screw profile 2)

Fig. 7.5 RVA patterns of corn-based extrudates produced from the DEEC-type process (from Clextral[10]).

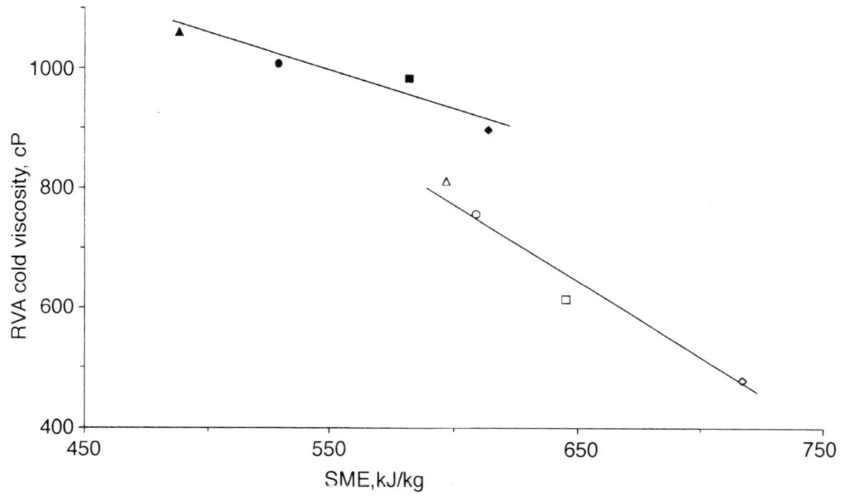

●: LS-HW-SP1; ◆: LS-LW-SP1; ▲: LS-HW-SP2; ■: LS-LW-SP2; ◇: HS-LW-SP1; ○: HS-HW-SP1; ❑: HS-LW-SP2; △: HS-HW-SP2.

(LS: low screw speed, 250 rpm; HS: high screw speed, 400 rpm; LW: low water content, 14%; HW: high water content: 17%; SP1: screw profile 1; SP2: screw profile 2)

Fig. 7.6 RVA cold viscosity as a function of SME (from Clextral[10]).

proportional to SME, and the slope of the linear relationship depends significantly upon the screw speed of the extruder; however it is apparently independent of moisture content and the design of the reverse screw element. RVA can also be used efficiently to investigate starch conversion of flake products made by PFEC-type processes.

Although the RVA methodology is an appropriate and promising way of describing and interprating the thermomechanical cooking of breakfast cereals, it needs really basic investigation to make the RVA signal more informative as regards the specificity of the extrusion-cooking process. It must be said that establishing straightforward correlations between process conditions and product attributes including sensory items is particularly difficult because current process and product parameters are not relevant enough. Therefore further study is still needed to complement instrumental methods and process methodologies in breakfast cereals processing.

Die texturization of directly expanded extrusion-cooked breakfast cereals
In the DEEC process, the cooking section brings the cereal melt to a rheological state compatible with satisfactory in-line die shaping. As a matter of fact, as the melt flows into the die, its pressure and temperature are about 60–160 bar and 150–180°C, respectively. In such conditions, any water is in the liquid state and perfectly mixed with the starch melt. When it emerges from the die, the cooked melt suddenly goes from high pressure to atmospheric pressure. This pressure drop causes an extensive flash-off of internal moisture and the resulting water vapour pressure, which is nucleated to form bubbles in the molten extrudate, causes the melt to expand. After expansion and subsequent cooling, the resulting product has a cellular structure the mechanical characteristics of which lead to a specific texture profile of the finished breakfast cereal product. In this process, a screw-barrel system cooks the cereal mix, and the die assembly must reveal the rheological potential of the cooked melt, so as to obtain the expected shape and texture of the product through die expansion and cutting. Thus, the expansion phenomenon is critical, as it relates cooking process history and product attributes.

At the final die insert, the cooked melt expands in two preferential directions as originally introduced by Alvarez-Martinez *et al.*:[11] the direction parallel to the flow which can be expressed by the Longitudinal Expansion Index (LEI), and the direction perpendicular to the flow which can be expressed by the Sectional Expansion Index (SEI). In fact, LEI is defined as the ratio of the exiting velocity of the extrudate after expansion, to the velocity of the melt in the insert; while SEI is defined as the ratio between the cross-sectional area of the product and the aperture area of the insert. The Volumetric Expansion Index (VEI) is then the product of LEI and SEI (VEI = LEI × SEI); VEI is, of course, inversely proportional to extrudate density, or product bulk density.

LEI and SEI values are particularly influenced by the extent of thermomechanical cooking and the design of the final die insert. Figure 7.7 shows the main process tendencies on an *expansion chart* (LEI–SEI diagram). At constant die design, but variable SME, increasing screw speed from 200 to 450

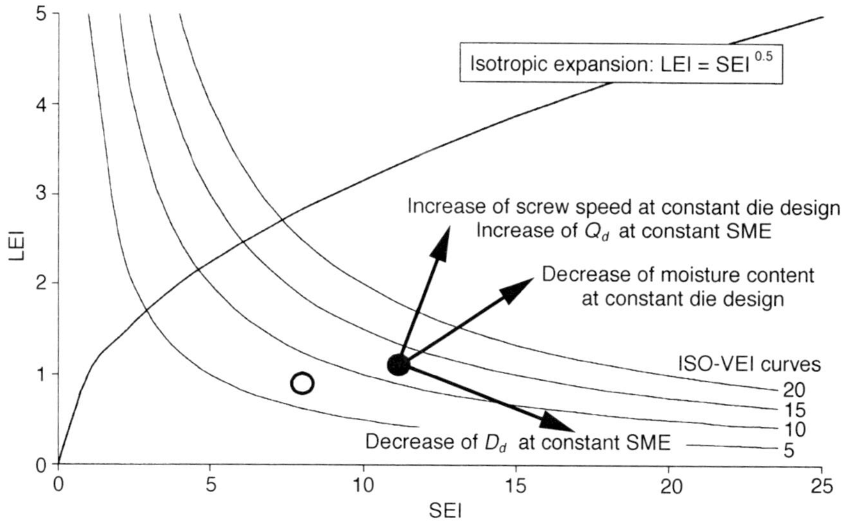

○: wheat-based recipes; ●: corn-based recipes; Q_d: melt flow rate per insert; D_d: insert diameter.

Fig. 7.7 Expansion chart of directly expanded extrudates. Effect of processing conditions.

rpm tends to increase substantially LEI and SEI; a similar result is also obtained by decreasing moisture content from 20 to 14%, but with a lower slope as compared to the screw speed effect. The screw configuration of the extruder and the design of shearing screw elements in the cooking section also significantly affect LEI and SEI. In general, high-time shearing elements in the cooking section lead to higher LEI and lower SEI; and inversely, when using high-shear rate elements in the cooking section. Although such changes in LEI and SEI are observed with all starchy materials, the expansion patterns are quantitatively influenced by cereal and starch types; for example, wheat-based recipes show lower LEI and SEI values than corn-based recipes (see Fig. 7.7), even though similar conditions of thermomechanical cooking are applied. At constant thermomechanical cooking (constant SME), but varying die texturisation (see Fig. 7.7), it is shown that both LEI and SEI increase as melt flow rate per insert (Q_d) increases; but LEI decreases and SEI increases significantly when die insert diameter (or equivalent insert diameter, D_d) decreases. The expansion phenomenon is generally anisotropic, and the practical domain in the expansion chart remains below the isotropic curve, as SEI is much more favoured than LEI. This was previously investigated by Bouvier *et al.*[12] Although the mechanism of melt expansion is not yet well established, the expansion anisotropy would be due to the melt elasticity. In fact, melt expansion is the more isotropic as SME increases[13] and shear rate in the final insert of the die decreases.[12]

Anisotropy of melt expansion at the die causes structural anisotropy of the resulting extrudates, with an impact on the mechanical properties and sensory attributes of directly expanded breakfast cereals. In fact, high SEI values, or conversely low LEI values, lead to relatively large cell sizes in the cell structure, with high sensory hardness and crispness of the extrudates. In fact, the expansion chart shows clearly that directly expanded extrudates may show variable cell structure (various sets of LEI and SEI values), at constant product density (constant VEI). The expansion chart is a useful tool for the product designer, to handle process and technological solutions which aim at varying cell structure and product properties through LEI and SEI parameters.

7.4.3 Drying and toasting

The role of drying in breakfast cereal processing is much more than product preservation through water removal. In fact, it aims to finalize the crispy/crunchy texture of products by reducing moisture content to the level at which cereal biopolymers are definitely in the glassy state. Moreover, drying may be combined with toasting so as to generate blistering and a specific brown colour and give a bakery taste to breakfast cereals, as in cornflakes and crisp rice processing. Drying must then handle time-temperature requirements for Maillard browning. In cornflakes processing, drying unit operation is more complex as it first creates the flake structure by air-induced puffing together with drying, and then toasts and cools the resulting puffed flakes. Thus, an engineering analysis of drying unit operation must include all breakfast cereal attributes such as texture, surface aspect and colour.

Predicting the drying time of puffed breakfast cereals

Puffed breakfast cereals like those obtained by the direct expansion extrusion-cooking process have low density foamed structures, with cell size ranging from 0.1 to 3–4 mm; bulk density of the products then ranges from 50 to 200 g/l. Generally, cell walls are thin, normally between 5 to 50 microns;[14] cell walls usually show a variable number of pores, which allow gas exchanges between contiguous cells. Based on the shapes of major breakfast cereals shown in Table 7.3, the characteristic length of the products ranges from 2 to 10 mm; characteristic length is the distance between the surface and the centre of the product, that is, half-thickness for a long slab sample, or radius for a spherical sample.

Drying usually occurs in a number of stages, each characterized by a different dehydration rate with hygroscopic materials like puffed starchy cereals. Figure 7.8 shows a typical desorption isotherm of puffed extrudate, which demonstrates the water activity domain where water becomes strongly bound to hydrophilic sites of the product (hydroxyl groups of carbohydrates, for example); such domain ranges from 0 to about 0.6 water activity, where drying occurs by desorption and then follows the second falling rate stage. When directly expanded extrusion-cooked breakfast cereals enter the drier, the water content is usually between 8 and 12%, which corresponds to the 0–0.6 water activity domain. The second falling rate drying process is normally controlled by internal resistance (water transfer controlled by diffusion in the extrudate), as

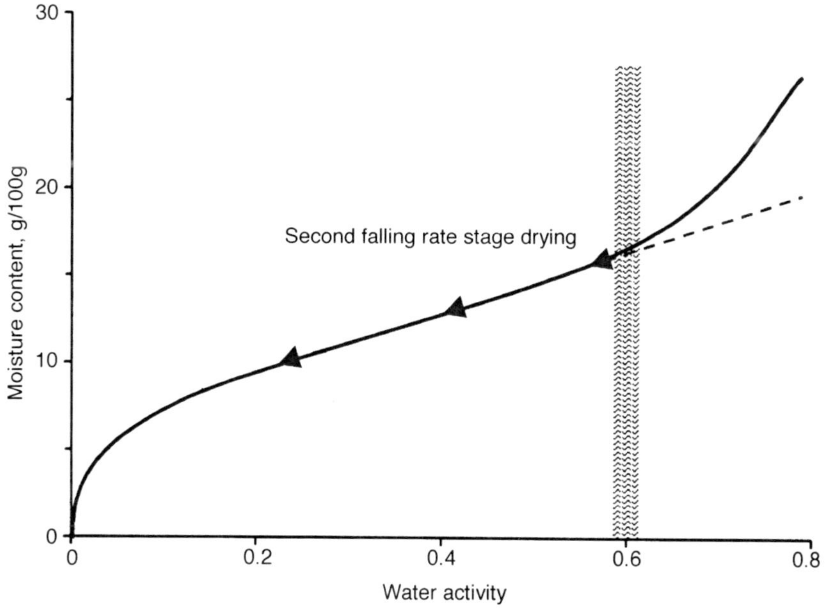

Fig. 7.8 Typical desorption isotherm of puffed breakfast cereals.

regards the high water transfer coefficient which characterizes the gas film surrounding the product. This makes it possible to calculate the time t_d required to bring the moisture content of the extrudate from the initial moisture content Mi, to the final moisture content Mf:[15]

$$t_d = \frac{4L^2}{\pi^2 D_e} \ln \frac{(M_i - M_e)}{(M_f - M_e)} \tag{7.2}$$

where L, D_e and M_e are the characteristic length of the extrudate, the effective diffusion coefficient for water vapour in the extrudate and the moisture content of the extrudate in equilibrium with the partial pressure of water in drying air, respectively. The problem of applying equation (7.2) lies in the lack of reliable data on D_e. In fact, D_e depends on the cell structure of the extrudate (cell size, cell wall porosity and cell thickness, in particular). As such cells are full of air, D_e would normally range between the water diffusivity in air (about 10^{-5} m^2/s) and the water diffusivity in starch (about 10^{-8} m^2/s), at the drying temperature: highly expanded products such as ball shape, loop and crisp rice, would probably show a D_e value of 10^{-6}–10^{-7} m^2/s; while dense products like bran-rich and oat-based products, would possibly have a D_e value of 10^{-7}–10^{-8} m^2/s. For a puffed breakfast cereal for which radius and initial moisture content are 6 mm and 10% respectively, and assuming that D_e equals 10^{-7} m^2/s and that M_e is negligible, the time required to dry the product down to 2.5% would be about 200 s (calculated using equation (7.2)); this time is actually the minimum dwell time of driers used in the DEEC process.

Drying process and driers used in the DEEC and PFEC processes

Industrial driers used in the DEEC process are designed to provide a wide range of dwell times so that they can cope with the diversity of drying times of puffed breakfast cereals. The flow behaviour of the driers must be close to plug flow so as to apply low dwell time dispersion, and thus obtain good homogeneity in the final moisture content of dried products.

The contact between the flowing air and the products must ensure a high water transfer coefficient from product to air (low external transfer resistance), and satisfactory air distribution through the products in a packed bed. The water transfer coefficient depends upon air velocity, which is at least 1.5 m/s as air flows upward or downward through the product bed.

Single and multiple conveyor driers are largely used for the DEEC process as they offer the required drying efficiency and process flexibility. They are equipped with various heating systems such as direct gas firing, steam heating coils and electric heating, in particular. Fluidized bed driers in which high velocity air tends to suspend products may also be used. Such technology allows very good air-to-product contact, and is particularly appropriate for toasting puffed cereals like crisp rice, where quick and homogeneous heating of the product is required.

In the PFEC process, flake puffing occurs together with drying. Air-induced puffing is actually the most important process step which requires a very high air temperature (240–260°C) and perfect contact between hot air and rubbery flakes, to obtain uniform heating and consistent puffing. In fact, each flake must be exposed to the same time-temperature history for reasons of quality, and then be completely surrounded by hot air. That is why the air-impingement drying technology, which fluidizes products, is the most appropriate system to texturize extrusion-cooked flakes satisfactorily.

Browning of extruded breakfast cereals and the toasting process

Non-enzymic browning involves the phenomenon of caramelization and/or the reaction of proteins or amines with carbohydrates, known as the Maillard reaction. Maillard reactions begin with condensation of a non-ionized amino group (ε-NH_2 of lysine or terminal α-NH_2) and a reducing sugar. Such reactions produce volatile components with desirable bakery flavours, and brown pigments which are responsible for the colour of toasted products. Adding sugars, milk powder and whey in breakfast cereal recipes then encourage Maillard browning when processing such recipes. Moreover, Maillard reactions have a high activation energy and therefore are markedly enhanced by high temperature heat treatments. It must also be mentioned that Maillard reaction rates are greatest at intermediate values of water activity (from 0.3 to 0.5).

Cereal toasting takes place at relatively high temperature (about 180°C), in the early stages of drying when the moisture content of products is compatible with high rates of Maillard browning. Then, air temperature decreases as the residence time in the drier increases, to avoid product overcooking and burning.

7.5 Future trends

Extrusion technology has modified substantially the processing environment of ready-to-eat breakfast cereals in the last three decades. In fact, extrusion technology allows product designers to process a wide range of recipes and raw materials, and to manufacture a wide diversity of shapes and textures, which fit in well with consumer demand. Besides, extrusion-cooking processes generally need short processing times, due to mechanical shear which efficiently cooks starchy materials, and lead to highly productive and flexible processes. All these characteristics contribute greatly to providing the consumers with various breakfast cereals at satisfactory cost.

As fas as extrusion-cooking is concerned, the future objectives that process designers and equipment manufacturers should concentrate on, would concern engineering and process developments:

- to improve both process productivity and product quality
- to develop new concepts of products.

The criteria that limit process output and product quality mainly relate to the lack of engineering knowledge about thermomechanical cooking of cereal recipes, and die texturization. For instance, the use of extrusion technology in flakes processing (the PFEC process) has made it possible to process a much wider range of recipes and raw materials, and to produce cheaper final flakes. It appears that better understanding of biopolymer conversion in twin-screw extruders would allow significant improvements in flake quality as regards the consumer demand for texture and taste in particular. Realistic extrusion-cooking solutions could be conceived and applied for that purpose. Die texturization is particularly critical in the DEEC process, and process performance is usually controlled by fluid flow in the die assembly together with melt expansion. In fact, a basic description of the melt flow mechanism in die systems would improve die scaling up and thus, contribute to optimizing process throughput and product consistency.

The breakfast cereal market increasingly reveals the need for new product concepts to accommodate health and nutrition requirements. Such a development should motivate process and equipment designers to develop appropriate unit operations required by market needs. Innovations to processes like co-extrusion, fruit- and seed-inclusion as well as pellet-based shaping, could fit in with future developments of breakfast cereals.

7.6 Sources of further information and advice

The author would just recommend these two interesting and relevant books to readers who want to focus more extensively on breakfast cereals:

FAST R B and CALDWELL E F (eds), *Breakfast Cereals and how they are made* St Paul, Minnesota, AACC, 1990.

KENT N L and EVERS A D, *Technology of Cereals*, 4th edn, Lancaster, Technomic Publishing Company, Inc., 1994.

7.7 References

1 EUROMONITOR, 'Breakfast cereals. Part 1: The breakfast cereals market', *Food Ingredients and Analysis International*, 1998 **19**(2) 19–21.

2 DATAMONITOR, 'Eastern opportunities for bread substitutes', *Food Ingredients and Analysis International*, 1998 **20**(3) 9.

3 DATAMONITOR, 'Latin America wakes up to breakfast cereals', *Food Ingredients and Analysis International*, 1998 **20**(3) 12.

4 CARSLAW H S and JAEGER J C, *Conduction of Heat in Solids*, New York, Oxford University Press, 1959.

5 LEVENSPIEL O, *Engineering Flow and Heat Exchange*, New York, Plenum Press, 1984.

6 KARATHANOS V T, VAGENAS G K and SARAVACOS G D, 'Water diffusivity in starches at high temperatures and pressures', *Biotechnol. Progress*, 1991 **7**(2) 178–84.

7 VAGENAS G K and KARATHANOS V T, 'Prediction of moisture diffusivity in granular materials, with special applications to foods', *Biotechnol. Progress*, 1991 **7**(5) 419–26.

8 BOUVIER J M, 'Engineering analysis of preconditioning in the extrusion-cooking process', *Cereal Foods World*, 1996 **41**(9) 738–40.

9 SMITH A C, 'Studies on the physical structure of starch-based materials in the extrusion cooking process', *Food Extrusion Science and Technology*, ed. by Kokini J L, Ho C T, Karwe M V, New York, Marcel Dekker Inc., 1992 **36** 573–618.

10 LEFEUVRE L, 'Détermination du degré de transformation de l'amidon dans les produits extrudés, au moyen du Rapid ViscoAnalyser', *Internal Research Report*, Firminy, Clextral, 1997.

11 ALVAREZ-MARTINEZ L, KONDURY K P and HARPER J M, 'A general model for expansion of extruded products', *J. Food Science*, 1988 **53**(2) 609–15.

12 BOUZAZA D, ARHALIASS A and BOUVIER J M, 'Die design and dough expansion in low moisture extrusion-cooking process', *J. Food Engineering*, 1996 **29** 139–52.

13 DESRUMAUX A, BOUVIER J M and BURRI J, 'Corn grits particle size and distribution effects on the characteristics of expanded extrudates', *J. Food Science*, 1998 **63**(5) 857–63.

14 GALLANT D J, BEN HDECH H and BOUCHET B, 'Influence de la cuisson-extrusion sur la microstructure de produits amylacés', *La cuisson-extrusion*, ed. by Colonna P et Della Valle G, Paris, Lavoisier, 1994 **5** 110–28.

15 KAREL M, 'Dehydration of foods', *Principles of Food Science. Part II Physical principles of food preservation*, ed. by Fennema O R, New York, Marcel Dekker Inc., 1975 **10** 309–57.

8

Snack foods

R. Guy, Campden and Chorleywood Food Research Association, Chipping Campden

8.1 Introduction

The name snack foods covers a wide range of food products. They are consumed as light meals or a partial replacement for a regular meal. Often they will be eaten while travelling or watching sports and other entertainments. The general range of snack foods will include products such as nuts, biscuits and merge into confectionery and meat products with count lines and jerky beef. However, the main sector, which is defined clearly as snack foods, contains the major snack products such as popcorn, potato chips or crisps and baked or fried snacks and starch-based snacks. There are many ingenious variations in the processes used by the industry, which serve to increase the range of products manufactured. A number of the most important processes are carried out using extrusion cookers as part of the production line.

The snack industry has been linked closely to the development and use of extrusion cookers for over sixty years. However, snack products predate extrusion cooking, having their origins in the traditional handmade products from earlier centuries. The classic examples of such products are puffed grains, prawn crackers and kerpok of the Far East and the tortilla snacks of South America.

It was possible to make prawn crackers in the home by cooking a 50% aqueous starch gel made from flour and fish-based bouillon to form a solid cake. This was sliced into thin pieces and dried in the sun to produce the first types of half-product. These stable intermediates could be stored until required and then transformed into a crispy expanded snack by frying in oil or heating in salt or even sand.

A second traditional technology grew up in the Americas where maize or Indian corn was a major cereal crop. The grains of maize were cooked in water until they softened and could be separated from their hulls and milled into dough. This dough was rolled into thin layers, cut into pieces and baked or fried to form tortilla chips.

These traditional products were dry and crispy and pleasant to eat and, with other products such as puffed grains or pork skins and nuts, formed the basis of early snack manufacture.

Extrusion cooking and ancillary processing machines have enabled these products to be taken from the cottage industry scale to be mass-produced at the rate of several tons per hour and to be made in many variations of form and recipes. At first the numbers and varieties of snacks may present a confusing picture to the food scientist and technologist trying to compare them and to understand their critical processing steps. For this reason we will attempt to classify the snacks into a number of categories based on the functional changes occurring to the raw materials as they are formed into products. For the modern industry there are several forms of snack products.

8.1.1 The major types of snack foods

1. Raw cut vegetable snacks

The fried potato slice has been developed into the best selling of all the snack products. It is based on thinly cut fresh potato, which is cut and washed before frying. This process dehydrates and creates a little expansion in the potato structure to form a golden coloured crispy product. Other vegetables and fruit such as carrots and apples have been used to extend the range of this type of fried snack with limited success.

2. Formed dough products from potato derivatives

The simplest snack products are made from doughs of dried potato derivatives and water. These doughs are mixed to a thick consistency formed by extrusion, or sheeting, and cut into small pieces such as tubes or discs. The moist dough pieces are passed into a frying bath to be puffed and dehydrated to form crispy products.

3. Formed dough products from maize derivatives

The second form of moist dough product is made from whole maize grains. These grains are cooked in water, washed and milled to form dough. The soft dough can be extruded, or sheeted, and cut to form moist pieces in the shape of flat strips or triangles. The pieces of moist dough may be fried or baked to form the crispy snacks.

4. Half-product or pellet snacks

The half-product snack is very similar to the formed dough in its origins. There are two routes to the product via precooked or raw materials. It may be prepared by mixing pregelatinised starch-based raw materials into dough. Alternatively, the process can also begin by cooking raw starch-based materials in an extrusion cooker to obtain the gelatinised starch dough, thereby simulating the method used in ancient cottage industries.

The important change in processing which distinguishes a half-product from a simple formed dough product is an extra drying step. After the forming process is completed either by extruding through a die, or sheeting and cutting, the doughs are dried slowly to a glassy state at 10–12% moisture. The half-product is a hard glassy material, which can be stored for a period of one year without loss of quality. In the final stage of snack formation the half-product is heated very quickly by frying or forced hot-air systems to puff. The expansion achieved in this puffing process is two to three times greater than the expansion achieved by dropping the moist doughs into a frying bath.

5. Directly expanded extruded snacks

The modern industrial snack was created in the early1940s with the manufacture of the first directly expanded snack from maize. In this process raw maize grits are fed into an extruder at low moisture to create a very hot melt within the barrel at temperatures of 140 to 180°C. It was found that a snack product could be created by releasing a continuous stream of the hot melt fluid from a small hole. As the pressure is released the melt stream generates water vapour and expands in microseconds to form a foam, which can be cut into portions by a rotating knife. The ribbon of foam is cut into short lengths of highly expanded crispy snack known as corn curls or puffs.

6. Popcorn and puffed wheat

The simple snack formed by heating popping maize in a saucepan is a highly expanded snack of greater specific volume than any other type of snack. A similar type of product can also be made from other cereals such as wheat and rice. Instead of utilising the strong hull of the grain as with popping maize the other cereals can be puffed by heating them under pressure in puffing guns or chambers. On firing the guns the water vapour inside the grains causes them to expand in much the same way as the popping corn to give a fine foam structure.

7. Related processes, such as snack biscuits and breadsticks

There are other snacks, which may borrow processing techniques from the main categories listed above or be similar in their mechanism. For example the simple dough products such as breadsticks, Melba toast and snack crackers are closely related products that rely on heating and dehydration to form their crunchy brittle structures.

In this chapter groups 1–4 above are the main subjects of interest although extrusion may be used in some instances of biscuit and breadstick manufacture. The chapter will be subdivided into sections to cover these groups.

8.1.2 Physical phenomena related to snack foods

The processes used to create snack foods are very ingenious and are easier to understand with the benefit of a basic description of the most important phenomena. There are three stages in the formation of all the snacks:

1. The formation of a dough by hydration of starch polymers so that they form a fluid mass that can be shaped into the individual snack pieces.
2. The heating of the dough mass in such as way that its water is superheated and released rapidly as vapour within the dough mass for puffing.
3. The snack is stabilised by being dried to low moisture levels to form a hard brittle structure.

Dough formation from starch-based raw materials requires that the starch granules are melted to remove all their crystalline structure. The physical form of the starch can be assessed by viscosity methods as shown in Fig. 8.1. In the initial melted form (B) they will hydrate and form a soft gelled mass, which can be mixed and sheared to form dough by the entanglement of some of the hydrated starch polymers as in D, E and F. This process can be carried out at low temperatures using pregelled starch or at high temperatures with native starch. In the latter case the starch is melted and then mixed and sheared at temperatures of 140–180°C.

Expansion of snack doughs occurs by releasing water vapour within their structure. This requires certain conditions to be set up in the dough. Ideally the water within a dough should be heated to well above the boiling point without

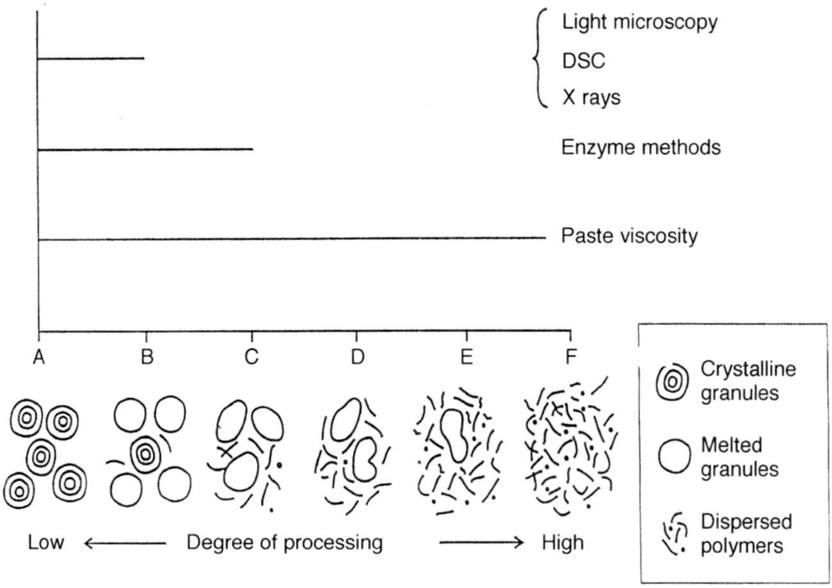

Fig. 8.1 Diagram showing the physical forms of starch and methods of assessment.

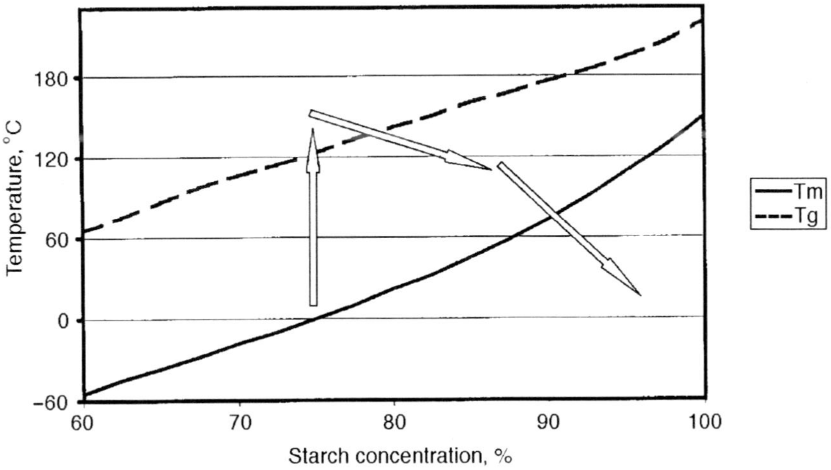

Fig. 8.2 Diagram showing the temperature of melting (T_m) and the glass transition (T_g) for wheat starch and the route which the starch follows during processing.

any evaporation until it all boils off. This can be achieved by rapid heating by dropping thin pieces of dough into hot oil or microwaving the samples. These methods create some expansion but tend to lose some water early and fail to create a large internal overpressure. The systems that confine the dough, such as extruder barrels or puffing guns, allow the water to be superheated up to 140–180°C without any loss. The superheated liquid water is vaporised in a fraction of a second on release from the equipment and forms bubbles throughout the dough mass. This can give a large expansion in starch-based doughs.

The final stage of creation for a snack involves the transformation of a soft dough, which has expanded to a foam, into a hard brittle structure. This process involves the cooling and removal of water from the expanded dough. Starch polymer systems are viscous fluids when well plasticised by water.[1] They increase in viscosity as the water is removed at around 8–10% moisture at ambient temperatures and as they cool they pass through the glass transition and become brittle and glassy (Fig. 8.2). This phenomenon has been studied in great detail in recent years[2, 3] and explains the hard brittle nature of all low moisture starch-based products such as snacks, breakfast cereals, toast, etc. The important conditions for the snacks to be glassy are low moisture and temperatures. If the snacks are heated to high temperatures they can become plastic or viscoelastic, but normally at ambient temperatures up to 40°C snacks with moisture contents $< 5\%$ w/w are crisp and crunchy.

These three simple phenomena are used in all types of snacks and it is the different ways in which they are applied that help to create the different forms of snacks available today. In the next sections of this chapter, we shall take a detailed look at the different forms of snack.

8.2 Formed dough products: potato

This group of products is made by extruding dough at relatively low temperatures to form basic shapes by forming or sheeting and cutting. The individual dough pieces are heated and dehydrated by frying in oil to obtain some expansion and to develop a crispy texture. The most successful products of this type have simple shapes, either as three-dimensional products such as sticks, tubes and hoops or as flat discs. The most technically sophisticated products are specially shaped discs known as stacking chips. In general these products are made entirely from potato derivatives but other materials have been included in certain products. The simple fried pasta type products used in Indian mixes also fall into this category.

The basic manufacturing processes for all these products are the same in principle but subtle variations are possible in ingredient selection and small variations of the processes, which give significant differences in product qualities. The processes include the following basic steps:

1. mixing dry ingredients
2. mixing water with dry ingredients to form a dough
3. forming the dough into product pieces by extrusion or sheeting and cutting
4. immediately frying for a few seconds to drive off the water
5. the products may also be dried a little more and tumbled, dusted or sprayed with flavouring.

8.2.1 The formation of dough

The formation of dough with suitable rheological properties for sheeting or extrusion and frying is the key to the process. This requires good selection of ingredients and the development of the optimum consistency in the dough by adjustment of water level and shear input to the dough to suit the raw material blend employed.

The main constituents of potato dough are dried pre-cooked potato derivatives. These powders can be mixed with water to form a sticky dough-like mass of mashed potato at moisture levels of 35–40%. Three types of material are available for use in potato doughs: potato granules, roller-dried potato flake and various forms of potato starches. The potato granules (Chapter 2) are dried to retain cellular structures in an add-back process. Dried granules are added to the freshly cooked wet mix to give a gentle equilibration to lower moisture levels and retain some of the structural integrity of the potato plant cells. The dried granules hydrate rapidly in an extruder and give more mechanical interaction in low moisture recipes, $<20\%$ w/w. However, they are designed for use in the high moisture recipes used to produce hoops and tube snacks from dough of 35–40% moisture. In these doughs the residual plant cell structure of the potato granules helps to provide a crunchier texture in the finished product. Some monoglyceride may be used as a processing aid in the manufacture of the granules and will be present in the material.

Potato flakes are formed by adding slurries of potato at 30% solids to the drum of a roller dryer, where it is heated to gelatinise the starch and flash off the water. This process tends to disrupt cells and granules to give a stickier product when the dry material is rehydrated than for the potato granules. Monoglyceride is added to help machinability on the rolls of the dryer, and is present in the material.

Potato starch is isolated easily from raw potatoes by cutting and washing the cells and may be modified chemically by substitution or cross-linking with a limited number of reagents. For 'Instant' materials they are subsequently gelatinised by roller drying.

In a standard recipe the bulk of the raw materials might be potato granules with 20–30% of flakes and 0–5% of starch. The materials will form a dough with 30–35% water w/w by mixing vigorously in specially designed mixers. The rheology of the dough can be manipulated from a moist crumbly structure to a more cohesive fluid by small changes in the moisture and shear inputs for a given recipe.

8.2.2 Shaping the dough

The dough is formed into shapes by one of two processes. Extrusion can be used to form continuous tubes or square extrusions that can be cut into short lengths at the die as individual product pieces. Well-known products such as potato tubes, sticks and hoops have been made and sold successfully for many years. In some products other cereals such as maize have been added to the potato. The extrusion of moist dough is fairly straightforward if the rheology has been adjusted to give good cohesiveness and flow characteristics. However, the processing window is fairly narrow for the basic recipe and special starches are often used to provide a wider margin for error. There will be some elastic recoil in the dough because pregelled starch has a large elastic modulus. The most important dimension of the extrusion is the thickness of the dough. This should be between 0.5 and 1 mm in a moist dough for a good texture to develop on frying.

The alternative to extrusion forming is sheeting and cutting which requires a good cohesive dough and may also suffer from elastic recoil after rolling out the sheet, changing the thickness.

8.2.3 Frying the dough pieces

The dough pieces should be fried at a high temperature of 185 to 190°C in vegetable oil (Fig. 8.3). At this temperature they change from dense dough to an aerated foam in 12–15 s and become yellow and then brown as they dry out. The critical factor in the frying is the thickness of the dough because if the dimensions are < 0.5 mm the dough may dry without aeration and if too thick it will aerate at the edges and remain moist in the centre after 15–20 s. Continuing the cooking of thick pieces will only cause excessive browning at the surface

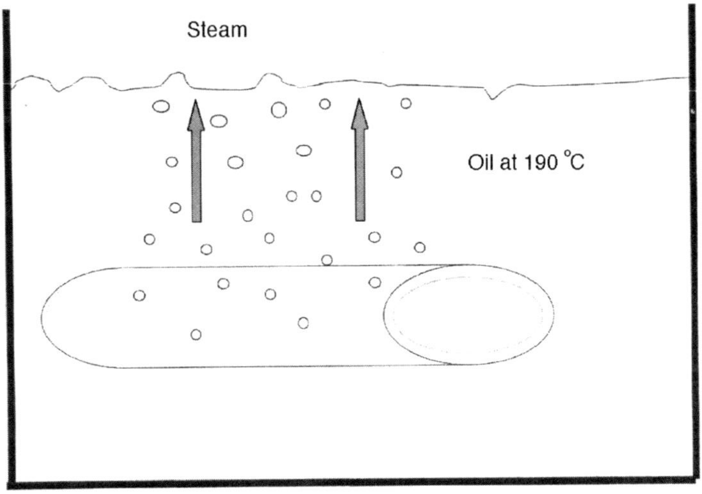

Fig. 8.3 Diagram of a dough tube frying to form an expanded tube showing the evaporation of water vapour through the hot oil.

while attempting to remove the required amount of moisture from the central region.

Drying the moist dough pieces a little before frying, which moves the process towards that of the half-product snack (section 8.4) can enhance the expansion of the dough pieces. However, the expansion is still fairly small and the products tend to be harder and denser than half-product snacks closer to potato crisps.

8.2.4 Variations in product types from formed potato doughs

There are two main types of formed potato dough products: the two-dimensional products cut from sheets of dough, with animal shapes or simple discs, and three-dimensional extruded products such as tubes, hoops and sticks. These products have been successful and taken their place as standard forms of snack products.

In recent years the attempts to manufacture a rival to the cut potato chips (crisps) has led to a very successful development known as the stacking chip. It was found that a chip could be formed from a potato dough by sheeting and cutting rough oval shapes but on frying these dough pieces did not match the appearance or texture of cut potato crisps. However, they had an attractive eating quality of their own, but were nondescript in appearance and easily broken into fragments.

The idea of forming the dough pieces in an attractive curved three-dimensional shape, so that every crisp was the same and could be stacked together solved both problems in one processing step. Dough pieces were sheeted and cut into oval shapes, which were placed in curved S-shaped moulds made of perforated metal. The moulds were passed through a hot oil bath, to dehydrate the doughs and form a hard crispy structure with a small amount of expansion. After depositing from the mould the crisps were stacked in cylindrical cardboard tubes that protected them from breakage before being consumed.

8.3 Formed dough products: maize and other materials

8.3.1 Basic processing from alkaline cooked grain

Moist dough products can be made from other raw materials such as cooked maize, wheat flour and gram flours. Maize products have been transformed to large scale manufacturing processes and are one of the most important forms of snack.[4, 5] The basic method for preparing the moist dough is to cook whole maize grains in water at temperatures up to 85–90°C[3] (Table 8.1). In most processes the maize grains are heated up from ambient temperature in limed water and after reaching the target temperature allowed to cool and stand for several hours to absorb the water and soften the husks. They are washed to remove hulls and excess of lime and then milled with stone burr mills to form a masa dough. This dough, which should contain 35–45% moisture, is formed from the shearing of some of the gelatinised starch granules and is similar to the potato doughs in its structure except that the starch granules are of different morphology.

The simplest processes will use the cooked maize doughs and form them into small pieces by sheeting or extrusion. The extruders, which may be either a ram or screw type, form the dough into sheets or ribbons, which are cut into simple geometric shapes such as triangles or rectangles. The moist dough pieces may be fried at 180–190°C in vegetable oil to form a crispy brittle snack with a small amount of expansion in the structure similar to the reformed potato crisp. Alternatively, the dough pieces may be baked in an oven at 300–315°C to form a denser hard brittle tortilla chip.

8.3.2 Precooked masa flours

The moist dough maize products can also be prepared from pregelatinised flour of maize that has been cooked to form masa dough and then dried to produce a flour. This flour contains gelatinised starch granules and can be reconstituted with water to reform the masa. However, it is possible to vary the texture of products by using combinations of different particle size masa flour. The use of finer particles is beneficial in baked tortilla chips, whereas coarse particles are necessary for the fried chip doughs.

The fine material takes up water more quickly and helps to form the cohesive glue whereas the larger particles are less cohesive, but form escape channels for the moisture during expansion. This prevents 'pillowing' and gives more even nucleation of bubbles within the doughs during expansion. There is a close parallel between the granular and roller-dried forms of potato and the larger and finer particle forms of masa flour.

Table 8.1 Plan for the manufacture of maize snacks from doughs

Operation	Material	Processing
1	Whole maize grain	water added (1.2 to 3 times grain mass) lime added 0.1 to 0.2% grain mass cooking at 80–100°C for 0.5 to 3 h steeping, 8–24 h
2	Nixtamal	washing to remove husks and excess lime milling in stone burr mill
3	Masa dough	drying grinding sieving
4	Masa flours	blending

In the cooked grains the dough characteristics may be controlled by the extent of starch swelling. This increases with water level in the cooked masa and gives more exudate the higher the water levels. In drier materials the dough will have too little of the sticky starch material which is liberated from the granules during cooking, and the dough may be short and crumbly. It is possible to add fine masa flour or an extruded maize flour to cooked masa doughs to supply more of the sticky cohesive materials and obtain a better sheeting quality.

8.3.3 Extrusion cooking of masa snacks

Attempts have been made to use extrusion cooking to produce tortilla chips directly in one straight run process. Such a process would take a few minutes instead of almost a day for the conventional products. However, the short time high shear extrusion cooking processes have been unsuccessful and a longer process has been required applying low shear extrusion. In a process described by Mapimpiant of Italy maize grains are presoaked in water for several hours and then mixed and conveyed to a low shear extruder. The extruder develops sufficiently high temperatures to melt the crystalline structures in the starch granules at 30–40% moisture and produce a dough roughly similar to the masa. This dough is sheeted using a second forming extruder. The dough was sheeted and cut and fried in the normal manner. In this process, the quality of the extruded snacks was claimed to be as good as the conventional one.

8.4 Half-product or pellet snacks

8.4.1 General concepts

The half-product method for manufacturing snacks is a two-step process. It requires the formation of a cooked dough of starch-based raw materials and careful drying to form a stable half-product. In the second step the half-product is heated rapidly in hot oil or air to achieve the expansion.

In the early processes the doughs were formed in a variety of ways but two main routes emerged. In the simpler method pregelatinised materials were mixed with water to develop doughs as described in section 8.2.1. The doughs were shaped into pieces and dried to a glassy state at 8–12% moisture to form the half-product. Many of these products were based on potato derivatives. In the second process method, mixtures of predominantly raw cereals were cooked in extruders to melt the crystalline regions of the granules and form a soft dough. This dough was shaped in dies or extruded as sheets to be cut into simple shapes. The moist dough pieces were dried to a glassy state to form the half-products. In both types of process there are critical features, which must be controlled for it to be successful.

- The starch granules should be melted to remove crystallinity but not dispersed more than about 10–15%.
- The moist dough must be dried carefully to avoid large moisture differences between the centre and the surface.

8.4.2 Extrusion cooking for the manufacture of half-products

For the first factor it is important to use low shear extrusion cooking in which the moisture levels are in the range 25–35%. The starch granules are more robust when melted at < 30% moisture because there is insufficient water for them to swell. However, the melting temperature increases with decreasing moisture and varies from about 115°C at 25% to 80°C at 35% moisture. The temperature is not usually a problem because there is sufficient internal heating to reach these temperatures. However, the control of the exit temperature is a problem for the single stage process (see Table 8.2).

A two-stage extrusion process is the simplest form of half-product manufacture because it is easier to control than a single-stage process. In the first extruder the conditions are set to melt the crystalline structures in the starch granules with minimum shear. The temperature may be raised to 100 to 120°C, or even higher with barrel heating to reduce the fluid viscosity and minimise the SME input. The hot melt fluid will expand at the die, flash-off some water and cool rapidly to 80–90°C. It then enters the feed port of the forming extruder and is compressed back into a dense fluid for extrusion through the special dies, or as a thin sheet. Therefore, it is possible to run the extruder at optimum moisture for extrusion cooking, reduce the moisture and cool the fluid in the flash-off for the forming process. This allows greater flexibility in the operation than for a one-stage process.

A single-stage extrusion requires a long barrelled extruder 25–30 L/D, with a cooking zone for the melting process near the feed input, followed by a cooling section downstream (Fig. 8.4). The starch must be melted in the first section and then the whole fluid cooled to < 100°C before it is extruded. It is important that

Table 8.2 Processing conditions for half-product manufacture with extruders

Process	One stage	Two stage
Machine	Long barrel twin-screw extruder	Short barrel twin-screw, or single-screw extruder
Zone 1	Reversing elements Temperature, 110–120°C	Reversing elements Temperature, 120–140°C
Zone 2	Forwarding elements Temperature, 90–100°C	None – fluid exits die
Forming extruder	None – fluid exits die at < 100°C	Compression single-screw at temperature 70–90°C

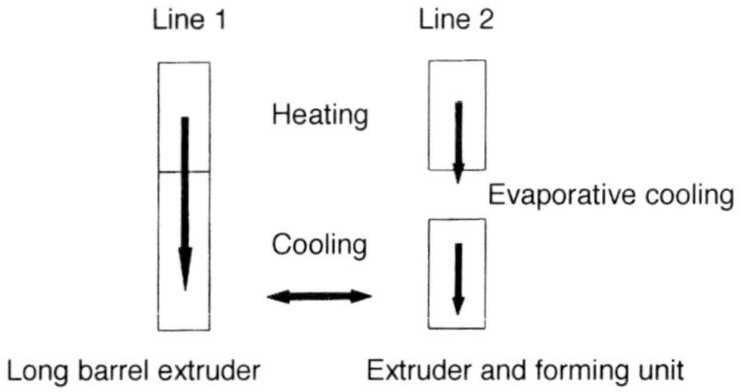

Fig. 8.4 Diagram of extrusion cooking lines for snack pellets (half-products)

half-product extrudates are free from any air bubbles. These would give distortions in the snack during expansion by forming large cavities. It is more difficult to control the cooking and cooling processes when they are linked together, and generally this will mean a reduced throughput on the unit used for the single-stage extrusion process.

8.4.3 Cutting and drying the half-products

The extrudate may be cut at the die face with a special cutting unit fixed to the die plate using thin flexible blades. They may also be extruded as ribbons or sheets to be cut later with a rotary cutter. The half-product pieces are usually cut as a thin layer < 1 mm thick.

After extrusion the half-products will have lost some water vapour and may have moisture contents of 20–25%. They must be dried further to form a starch glass at around 10% moisture. This is not a simple process because moisture evaporates from the surface faster than it diffuses through the gelled starch. Rapid air-drying at high temperatures forms hard dry skins on the half-products with moist centres. This will set up strains within the structures, leading to cracking and also lead to pillowing and the formation of large cavities during expansion.

In order to obtain a small moisture gradient across the half product, drying is carried out at 50 to 60°C at a relative humidity that reduces from 65 to 45% over

6–7 hours. After completing such a drying process the half-products have moisture contents around 10–12% and are brittle in texture. They can be expanded into light foam structures at any time for up to at least one year.

8.4.4 Expansion of half-products

Half-products have a dense glassy structure that contains a small amount of water. If it is heated very quickly, so that its temperature rises fairly uniformly across the individual piece, the water will become superheated within the glass. The polymer glass will pass through its transition to become a viscous fluid at a very high temperature at 10% moisture. This temperature will be in the range 130–150ºC and well above the boiling point of water. As the half-product structure is heated above this temperature and becomes fluid, it will flow under the high internal pressure generated by the water vapour and bubbles will form and expand with the fluid matrix.

The nucleation of the bubbles is very important to the final texture because the more nucleation the finer the texture. It was shown that nucleation is related to very small bubbles already present in the glassy structures. These have been shown to be retained from the structures of the starch granules.[6] If the starch granules are highly dispersed in the extrusion process the nucleation is greatly reduced and the snacks are coarsely textured. This appears to be the reason for the success of the low shear extrusion processes. It is possible using the starch pasting methods, which are easily performed on a Newport's Rapid Visco Analyser, to measure the physical state of the starch and to optimise the processing for nucleation.

8.4.5 Three-dimensional half-product snacks

The normal half-product snack is formed in one piece and expands from this piece into a small flat snack. In the latest innovations two layers of unexpanded extrudate are combined by cutting, or punching, shapes from the combined sheets to form a two-layer half-product. This is dried in the normal manner to form a shelf stable intermediate. On frying in hot oil or hot air the half-product expands in two ways. The individual layers expand as finely textured snacks and the two inner surfaces are pushed apart by a single large bubble, which forms between them. Thus the snacks forms a pillow or cushion shape with textured layers encasing a hollow centre. These snacks are similar to co-extruded pillows with no fillings made by direct expansion but have a finer texture and better sensory properties.

8.5 Directly expanded snacks

The directly expanded maize snack was the first industrial form of snack food. The Adams Company manufactured it in the 1940s from maize grits using a short-barrelled single-screw extruder. It continued to be manufactured for many years before the principles by which it was produced began to be understood and indeed it was not until the 1970s and 1980s that the true nature of extrusion cooking for direct expansion was determined by careful research.

8.5.1 The basic principles of direct snack extrusion

The transformation of starch-rich feedstocks such as maize grits, wheat, rice or potato flour into hot melt fluids, which can be expanded as they emerge from a die, occurs on the screw between the feed port and the die. In the early machine this area was inaccessible to the scientist. At CCFRA in 1983[7] it became possible to observe the changes in raw materials using a split barrel twin-screw extruder. The first studies examined the development of wheat flour and maize grits in the screws under the ideal running conditions to produce extruded snacks. It was also possible to vary the processing parameters, including the screw sections, and to follow the dynamic process variables and measure the product characteristics. The process was examined on a standard screw system comprised of three main sections, conveying and mixing, back pumping and shearing, pumping for extrusion at the die. This simple system was adopted after studies with more complex screw designs were found to be no better and often less stable.[1]

Examination of a 'dead-stop' shutdown showed that in the conveying section the screws were partly filled and the material was not compressed or sheared. Almost all the heat was obtained from the barrel and the temperature rose to 70–80°C for a barrel profile for three sections 2.5 L/D of 30/50/90°C. The heat transfer was related to particle size and smaller particles heated up more quickly.

The reversing or back pumping section caused the feedstock to fill some of the conveying screws completely. This allowed the screws to pump the material forward against the reversing section until the forward pressure overcame their resistance and set up a steady forward flow of material. The powder filling the forward pumping screws was compressed to about 1 ml/g and filled the reversing paddle elements. In the first paddle pair the pressure was 20–30 bar and the particles were sheared against each other and the metal elements. This caused a large input of heat within a distance of a few mm and a rapid increase in temperature. For the normal low moisture levels used in direct snack extrusion of 16–18% moisture, the temperature rose from 80 to 150°C in the first paddle (12.5 mm along the shaft).

Examination of a maize or wheat flour feedstock showed that the crystalline granules of the starch became amorphous and were squashed in the first pair of paddles in the reversing section. The soft granules were deformed to flat pancake shapes and then dispersed into a starch polymer continuum. Further examination of the starch within the melted and sheared flour showed that the starch polymers themselves had been degraded. Amylopectin (AM), which occurs naturally as a very large branched polymer, was reduced from $10^8 D$ to $< 10^6 D$, during a snack extrusion. The material within the reversing section was

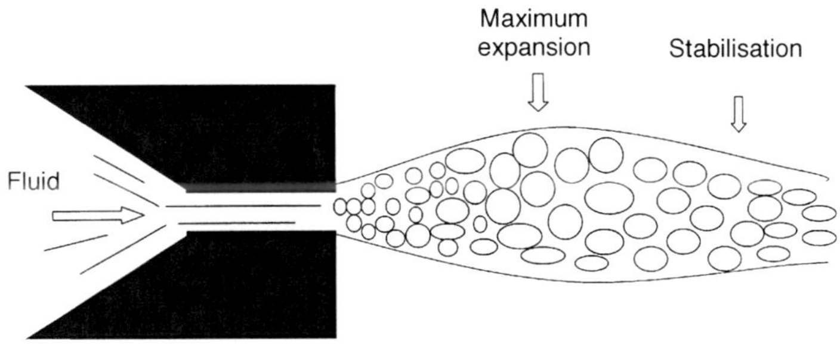

Fig. 8.5 Diagram of the expansion of an extrudate at the die of an extruder showing bubble growth and stabilisation of the foam.

transformed from a compressed powder at the start of the section to a viscous fluid at the end. Cooled samples of the fluid were found to have a density of 1.4 g/ml and contained no visible air bubbles. Measurements of the hot melt fluid in a special die showed that it had a viscosity of 3–4 kPa.s.

The internal temperature of this melt fluid was in the range 140–180°C depending on the barrel heating or cooling applied. At these temperatures the water was present as superheated liquid contained within the barrel at pressures of 40–80 atmospheres. Extrusion of the hot fluid out of the machine lowered the pressure to atmospheric and caused the water to vaporise within the fluid. This high-energy process enabled bubbles to be created and expanded within the melt to form continuous foam. The expansion continued until the cell walls reached the limit of their extensibility and ruptured releasing the gas pressure. It was shown that the limiting factor in expansion at these low moistures was the cell wall extensibility and not the gas pressure. After the gas cells ruptured (Fig. 8.5) the expansion ceased and the extrudates began to contract under elastic recoil but were stabilised by the rapid temperature drop viscosity increase accompanying the evaporation of up to 8–10 g of moisture per 100 g of fluid.

The basic structure within these snacks is formed by a starch continuum. Other materials are dispersed within the starch. As the starch polymer system dehydrates and cools it approaches the glass transitition. This varies with moisture but is at 40–60°C for snacks of around 5–8% moisture. After passing through the glass transition the texture of the snacks becomes brittle and provides the normal sensory characteristics of a snack.

8.5.2 Fine details of snack formation

The extrusion process concentrates on the transformation of the starch within the recipe but may be modified to produce a range of snacks with different characters. This may be done by changing the recipe and also by changing processing conditions.

Flavouring

The recipes are based on major starch-rich ingredients such as maize, rice, wheat or potato. These materials are different in terms of their composition and their minor ingredients such as protein, lipid, carbohydrates and fibres cause small differences in their physical performance in the extruder and significant differences in flavour and colour. It is possible to blend any of the materials together to form a starch continuum, which will form an expanded foam structure. If the starch level is maintained at a constant level it is possible by adjusting the processing inputs to manufacture similar expanded snack structures. However, the colours of the major raw materials mentioned above are sufficiently different to cause significant differences in extrudate colour. These may be adjusted to consistent levels in certain colour ranges by adding colours to the recipe.

The flavours generated in snacks by the basic raw materials such as maize, potato, rice and wheat are also different in character and intensity. They are also changed by processing in relation to temperature and time spent in the hot zones of the extruder.[8] It is possible to adjust the flavour of snacks by coating them with added flavouring in post-extrusion processes.[8] However, the contribution to the overall flavour from the background flavour within the snack is significant and helps to determine the consumers' preferences.

Studies on the flavours generated within the extruder from the basic raw materials have been supported by other studies on the effects of adding flavour precursors to the raw materials.[3,9] These are usually a combination of small active materials such as reducing sugars and amino acids and peptides. At the high temperatures within the extruder they undergo Maillard reactions and other thermo-chemical reactions to produce new flavours and colours. Therefore, in any recipe for extrusion cooking attention must be paid to all the ingredients added. They may contain small amounts of materials that are active in these chemical reactions and may change their nature. For example, if milk solids are added they will introduce a reducing sugar lactose and a mixture of proteins and peptides. These will undergo Maillard reactions and add flavour and red-brown colouration to the extrudates.

Texture modification

Pure starch polymers will form finely textured foams in extrusion. Their cell walls will be extended to give very thin dimensions and the fluid may be manipulated to give a small number of large cells or a large number of very small cells. The formation of bubbles in the starch melt is affected by the presence of nucleating substances, viscosity and the manipulation of the die pressure.

If finely divided insoluble materials such as calcium carbonate (1–50 μm) are added at 0–2% by weight of the starch the numbers of gas cells observed in the extrudates increases from a few hundred to 70,000 ml^{-1}. Several types of materials have been found to have the same effect. The similarity between them is their size range and the fact that they are insoluble in water at temperatures of

100–180°C. Inorganic salts have the most powerful effect but the addition of insoluble fibres such as bran from wheat or other cereals and the hemi-celluloses in barley are less effective and need to be present at higher levels of 3–10% by weight of the starch to show significant changes in texture.

Shaping and cutting directly expanded snacks

The shape of a snack is partly determined by the die, particularly its cross-section and the cutting action of the knives. For a starch-based extrudate with a basic recipe of predominantly cereal or potato flour, the fluid melt is viscoelastic with a high viscosity value and a strong elastic modulus.[5] This means that it will have die swell if it is pumped through a small die channel of a few mm maximum dimension after being conveyed through barrel sections of 65, 90 or 100 mm diameter. For simple circular dies the die swell will give increased radial expansion at the expense of the linear expansion. This is not a problem until small diameter snacks are required and the die holes have to be very small and can easily be blocked.

For more complex shapes such as animals or squares the die swell effect must be allowed for in the die design. Usually this is achieved by trial and error because the values for the rheological characteristics are not known with sufficient accuracy for mathematical modelling. However, advances are being made and there are some companies which offer modelling services for die design.

There are two factors, which reduce the effect of die swell and allow the shape of the extrudate to follow the die cross-section more closely:

1. Addition of a filler such as bran or fibrous materials which dissipate the stored elastic energy within the fluid as it flows in the die and emerges from the die.
2. Addition of nucleating substances which form a large bubble population for expansion and a fine texture. Small bubbles dissipate the elastic energy in all directions rather than at right angles to the direction of flow and therefore give an isotropic form of expansion.

Therefore certain recipes are easier to shape into more precise designs.

Some novel product shapes have appeared on the market, which rely on the application of new cutting technology. Three-dimensional shapes of animals such as horses and dinosaurs have appeared, which have four well-defined legs. This new shape was achieved by using two cuts from two knives. The first knife cuts the full area of the animal while the second blade following close behind only cuts across part of the collet. The second cut divides the legs of the animal increasing them from two to four.

8.6 Co-extruded snacks

The idea of running two extrusions through a single die is well known in the plastics industry. It is much more difficult when operating with food materials because of the large changes in viscoelastic characteristics, which can occur in starch-based fluids. However, the extrusion technologists have achieved some remarkable products by co-extrusion in both the human food and petfood markets.

8.6.1 Co-extrusion of a biscuit with a soft filling

The co-extrusion of an expanded cereal biscuit in the form of a tube or flatbread with a soft savoury or sweet confectionery filling was brought successfully to the market by several companies. Biscuit recipes such as those in Table 8.3 have been developed using the normal process for an expanded snack and extruded through an annular die. The expanding biscuit forms a tube around the central tube. A small projection of the tube for 1–2 cm or so from the die allows the tube to stabilise with a hole the same size as the tube.

The biscuit recipe can be varied to utilise any of the major cereal types or potato flours. In our studies a blend of wheat flour and potato starch was found to give a light finely textured biscuit. The fine texture was judged by consumer trials to be preferred to coarser textures for snack tubes. They also gave a smooth inner surface, which prevented leakage into the biscuit of the hot filling. The biscuit may be extruded with a filling to achieve a moisture content of 5–6% and an ERH% of 45–50%.

A special fat-based filling is used with a recipe as shown in Table 8.3. This is based on a fat system designed to melt at 30–35°C and to give an ERH% equivalent to or less than the biscuit. The filling must be fluid to pump into the extrusion die but must set to a solid form quickly while the filled tube moves from the die head to the cutting station. It must also have good sensory qualities, which means that all the fat must melt in the mouth so that there is no greasy mouthfeel. Therefore a fat blend with a melting point of 30–35°C such as hardened soya or palm kernel oil is required. As it is pumped through the die tube to emerge inside the hollow cylinder of biscuit, its temperature may rise a

Table 8.3 Recipes for biscuits and fillings

Ingredient	Biscuit	Savoury filling	Sweet filling
Flour	60	33.2	24.7
Sugar	10	8.3	20.6
Soya isolate	5	4.2	0.0
Potato starch	15	8.3	8.2
Skim milk powder	3	5.8	5.8
Whole milk powder	1	0.9	0.8
Salt	1.5	1.2	1.2
Calcium carbonate powder	0.75	0.0	0.0
Oil and fats, hardened palm kernel oil	1	29.1	28.8
Flavouring	0	4.2	4.1
Water	2.75	4.8	5.8

few degrees, because the fluid surrounding the tube is at 130–150°C. However, with a high flow rate in the small bore tubing, this rise is fairly small and the fat can be pumped into the tube at 38–40°C. This allows the fat to reset rapidly over a setting distance of 10–15 m.

The rates of flow of biscuit and filling are adjusted so that the degree of filling of the biscuit is at the required level. In a simple filled cylinder the extrudate is conveyed over a distance of 20–30 m before being cut into short lengths with a guillotine cutter. During the conveying time the filling sets to a pasty solid so that it remains in the tubes and gives a clean cut. The moisture content of the tubes must be <7% w/w for the product to have and maintain a crunchy texture during storage as a normal snack food product.

Several products have been manufactured in which the extrudate is only partially filled and is crimped into small lengths within a few metres after extrusion while the biscuit is still hot and soft. In this case the filling is trapped within the biscuit and cannot leak out. The biscuit is compressed into a seal at each end of the pillow and then sets to a crisp texture on cooling to ambient temperature. It is possible to dry this type of product after extrusion without losing any filling from the biscuits. Therefore, such products are more suitable for hot climates than the open tubes because the fillings can be made with a melting point of <35°C so that they do not taste greasy in the mouth.

8.6.2 Co-extrusion with secondary filling

There are several problems in using a filling that passes through a hot die head system. The heating is not a real problem when running smoothly but the filling can burn on to the tubing if there are any stoppages in the filling line or shutdowns on the extruder. This leads to blockages and the clean-up procedures are laborious. An alternative system was developed in which an empty biscuit extrusion was filled downstream from the extruder with one or more fillings using injection nozzles. An open U-shaped extrusion was filled by injection of one or more fillings into the trough formed by the tube within 200–300 mm from the die. Immediately after the injection the soft plastic textured tube was compressed in a sealing device to close the gap and form a sealed tube.

This type of filling system could be used to form short length or sealed pillow products as in section 8.6.1. However, the same rules of sensory and moisture activity apply to these fillings. They must have a fat system, which melts in the mouth and an ERH% value less than the biscuit.

8.6.3 Co-extrusion of two biscuits

The manufacture of a co-extrusion of two expanded starch-based foams to form snacks requires the use of two extruders as in the case of plastics extrusion. The viscosity of a fluid melt of starch is too high for normal pumping systems and for pumping through any long sections of tubing. Therefore the best systems require two extrusion cookers to be set up to pump their fluids in a single die head. The possible forms of mixed foams are determined by the geometry of the dies, which must allow both streams to expand. The inner stream would need to be extruded into a wider tube for expansion and the outer stream would be extruded in an annulus around the inner tube system. This type of product has been made for the petfoods market with fairly dense extrudates but has not been exploited successfully in the snack market.

8.7 Future snack processes

8.7.1 Innovations

The snack manufacturer may use extrusion cooking or extrusion forming in many ways allied to rapid heating technologies. Many ingenious patents appear every year with new ideas for combining these processes in new pieces of equipment or using them on new raw material combinations. The main directions being taken in the snack market have not changed very much, but some influences come from the general changes in the food market and others from the media. The Internet will play an increasingly important role in spreading ideas for products and processes.

8.7.2 Quality aspects

The sales of snacks to adult markets have led to improvements in eating quality of snacks with respect to sensory texture and flavour. The simple maize puffs flavoured with a dusting of powdered flavouring and oil have less appeal for adults than children. Therefore the recipes and processing conditions have been changed to increase density, cell wall thickness and crunchiness and reduce the stickiness and tooth packing problems of simple snacks.[10]

Some manufacturers have adopted the use of water-based flavouring to replace the high oil levels added to snacks. These use a spray of aqueous hydrocolloid solution to carry the flavourings and a final short drying process to reduce the moisture to the normal range $<5\%$ by weight. Oil levels can be reduced to about 15% w/w without losing the sensory appeal but lower levels tend to bring adverse comments and rejection.

The addition of proteins and fibrous materials to snacks at levels of up to 20–30% has been explored to improve both the textural and nutritional qualities of snack foods. These ingredients require some thought and the recipes and processes may need to be modified to accommodate them, but when minor difficulties are overcome the products are of a higher quality for the adult market.

One interesting product, which has been manufactured at CCFRA, was made from pea flour, which contained all three ingredients required for a good snack texture, starch, protein and fibre. This raw material formed some excellent snacks with an attractive pea flavour.

8.8 References

1. GUY, R.C.E. and HORNE, A.W. (1987) 'Extrusion and co-extrusion of cereals', Chapter 18, in *Food Structure – its creation and evaluation*, eds J.M.V. Blanshard and J.R. Mitchel, pp. 331–50, Butterworths, London.
2. ROBERTS, S.A. and GUY, R.C.E. (1986) 'Instabilities in an extrusion cooker: a simple model'. *J. Food Engineering*, 5(1), 7–30.
3. BREDIE, W.L.P., MOTTRAM, D.S. and GUY, R.C.E. (1998) 'Aroma volatiles generated during extrusion cooking of maize flour', *J. Agricultural and Food Chemistry*, 46, April, 1479–87.
4. SERNA-SALDIVAR, S.O., GOMEZ, M.H. and ROONEY, L.W. (1992) *Advances in Cereal Science and Technol. Volume X*, Chapter 4, 243–302, American Association of Cereal Chemists, St. Paul, Minnesota.
5. GUY, R.C.E. (1995) 'Breakfast cereals and snackfoods', Chapter 18 in *Physico-chemical Aspects of Food Processing*, ed. Beckett, S.T., pp 368–86, Blackie Academic & Professional, London.
6. MATZ, S.A. (1993) *Snack Food Technology*, 3rd edn, AVI Publishing Co. Inc., Westport, Connecticut.
7. LAI, C.S., GUETZLAY, J. and HOSENEY, C. (1989) 'Effects of baking powder on extrudates'. *Cereal Chemistry*, 66, 69–73.
8. HOSENEY, R.C., ZELLEZNAK, K. and ABDELRAHMAN, A. (1983) 'Mechanism of popcorn popping'. *J. Cereal Science*. 1(1), 43–52.
9. BREDIE, W.L.P., HASSELL, G.M., GUY, R.C.E., MOTTRAM, D.S. (1997) 'Aroma characteristics of extruded wheat flour and wheat starch containing added cysteine and reducing sugars'. *J. Cereal Sci.*, 25, Jan., 57–63.
10. GUY, R.C.E. (1999) 'Tackling snack food formulations'. *Savoury & Snack Foods Industry Guide* (Europe) 18–20.

9

Baby foods

M. Kazemzadeh, Buhler Inc., Bloomington

9.1 Introduction

Use of the extrusion process in the food industry has taken a different turn in recent years. New concepts, products and applications are appearing in the market as the industry masters this technique of food processing, and the concomitant advancements in engineering and processing. One of the more unique applications of extrusion has been the application of a low-shear-high-pressure and residence-time extrusion technique in the manufacture of baby foods and toddler snacks.

In order to appreciate fully the various benefits of extrusion, it is essential to look at the traditional process of cooking and compare it to the new extrusion process. Not until the researcher understands the limitations along with the benefits of the extrusion process will he or she be able to manipulate the parameters that will best suit its purpose for the manufacturing of specific products. In examining the applications of extrusion in the food industry, it is also essential to study the range of limitations set forth both by the various designs of extruders as well as the extrusion processes presently in the market.

The field of extruded baby and toddler foods, especially in the organic and natural foods market, is growing rapidly in both the US and Europe. This growth will drive the direction the food industry takes in this field. The trend is toward more nutritious and wholesome foods that can be produced inexpensively. Investors, business, governments and funding institutions will want to observe this trend as they consider the potential for future markets. The application of this process along with innovative developments will have a profound effect on the social-economic nature of the future market. This will place a major responsibility on the shoulders of the food companies to fill the gap in the consumer market by providing high quality, nutritious, and delicious foods. The ease and lowering of the cost of production of such products, coupled with the freedom to use any source of raw materials locally available, e.g., rice, corn, various legumes and fruits, makes the extrusion cooking process an attractive investment in the area of food production.

Finally this process lends itself well to the implementation of resources earmarked for feeding the world population. Utilization of this technique will bring a better-tasting product to the baby and toddler market. Further, it can produce foods designed for teens and adults with a complete nutritional profile that is also shelf stable.

9.2 Traditional batch processing

While there have been great advances in the field of food processing and food engineering in the both the European and US markets, many aspects have been left unchanged. Food manufacturers continue to use traditional processing techniques in manufacturing products to meet the demands of the marketplace today. The traditional processing of baby foods is a prime example. Baby/toddler foods have become convenience foods. The major characteristic of a good baby food is its ability to hydrate readily, be fully gelatinized and cooked by simply adding hot, warm or tap water to make it readily consumable. These foods tend to be starch-based products. One way to enhance these foods is to insure that they are fully cooked and prepared in a manner that does the least damage to the starch granule during processing. This will result in more viscous and full-bodied finished product, which consumers tend to prefer.

Two such processes will be reviewed in this chapter; the first is the traditional pre-gelling process. This process requires the complete hydration of the finely ground flour. The drum drying technique is then used to gelatinize the starch, slightly modify the protein portion of the grain, and package the product in flake-like format. Low bulk density and easily re-hydratable flakes characterize such products. The product can then be mixed post cooking, with other sweeteners and ingredients to produce a final dry meal. The addition of water creates the desired gummy consistency. This gummy portion can then be heated or served warm or cold to children of young age.

The main drawback of this process is the amount of energy needed to grind and hydrate the grain and dry the finished product. The drum dryers are bulky in size, difficult to keep clean, and inefficient in their heat energy transfer. The addition of many non-starch products to the recipe, e.g. sweeteners, tends to make the process extremely sticky. This results in the product itself adhering to the surface of the drum, thus making the manufacture of recipes using this process difficult, unproductive, and impractical. Further, they are usually limited to a low production rate of 800–1000 lb/h.

The main benefit is that this process insures that the starch granules are not damaged during the heating and gelling process, thus resulting in the highest viscosity for the final product. The process requires the starch slurry to be hydrated to about 75–85% before gelatinization on the drum dryer. The same amount of water must then be removed during drying. Although this process is continuous in its nature, inconsistency in the product, due to such factors as variation of temperature in the heating of the drum, grinding size, and quality of water for hydration cannot be detected until a large amount of the product has been processed.

The aforementioned limitations make this a batch process rather than a true continuous system. The large water requirement for this system, along with the waste water generated from this process, can be limited by government regulations. This can further limit production ability of the total factory depending on the site of production. Finally, to reduce the grain size from full kernel to powdery flour, a liquid or dry grinding technique can be used. The grinding techniques necessary further limit the production rate of such a system.

The second process for making traditional products is the use of batch cookers where the total grain in various sizes and grind form is added to a heated chamber with additional flavoring, sweeteners and water to be hydrated while the total chamber is heated and mixed continuously. These cookers are used widely in the cereal industry and are considered still to be one of the best methods to produce a cereal which has the least amount of damage made to the grain starch. It uses grids or larger grain particulate for its process. Since there is no shear in the system it tends to generate the least amount of starch fragments or dextrins. Such a system is also considered to be outdated due to its enormous problem with cleaning and its labor-intensive operating technique. After the product has been cooked in such chambers it is then dried and flaked or ground to smaller size particulate and then packaged for sale. Such processes yield excellent products with a complete flavor profile and a high viscosity final product after hydration at the consumer level. Major problems associated with such processes have been the lack of ability to clean the system completely after each shift, and the extensive labor, space, and waste water generated during cleaning that limits the success of such processes in this day and age. These systems are still in great demand and with a well-managed factory can produce a good finished product for various markets.

9.3 Extrusion system for baby foods

The introduction of a cooking extrusion system into the food industry, initially in the early 1960s, triggered a large opposition from manufacturers using the traditional process. The system was continuous and easily manageable for factories of various sizes. Its first introduction was into the petfood market followed by the cereal and other markets. Its basic processing technique was to generate heat by mechanical energy or the shear rate generated by the turning of the extruder screw against the barrel. This process produced high temperatures by rapid grinding of the grain within a closed and pressurized system. The further development of the extrusion process led to the development of the twin-screw extruder which utilized the development of the single-screw extruder's conditioning process followed by the use of heat exchange capability of the extruders to bring the product to high temperatures under more controlled parameters and less dependency of grinding technique within the extruder to generate the heat required to cook the product.

9.3.1 Defining cooking of cereals

The term 'cooked' or 'cooking' is an ambiguous term used commonly in the market to describe a process which can be defined scientifically for a given ingredient such as a starch granule's evolution through a glass transition phase, where the starch granule finally can be considered gelatinized and the protein globules are textured. Since most of the ingredients for baby foods are starch- or sugar-based we will try to explain the terminology that pertains to these portions of ingredients. Starch is usually in a granular format in its native stage, and may vary in size from a few microns in diameter for small grains to a few hundred microns for tubular starch source such as potato. Gelatinization of starch granules that may contain from thousands to millions of starch molecules is indicated by the disappearance of the Maltesian Cross that is detected under a light transition microscope with polarized filters. The native starch granule depicting such characteristics under a light transition microscope during gelatinization begins to fade out its distinct granular boundary as the granule is hydrated and heated. This is a process by which the granule will lead to full gelatinization or disappearance of the Maltesian Cross at a given temperature and moisture level. Different starch granules reach this stage at different temperatures in the presence of ample moisture.

The hydration and gelatinization of the starch granule at a given constant temperature of, for example, 180°F in a petrie dish, will have an evolution of three various stages. The first stage is the swelling of granules followed by a total loss of ridge identity, and finally bursting and dissolving of the boundary lines of the granule whereby the starch molecules will begin to interact with each other, and form a continuous matrix. Once the matrix is formed, the mass develops a viscosity much higher than its original mixture of water and starch granule. It is the viscosity of the above dough within the extruder that provides the utilization of convertive or mechanical energy into heat energy. In other words, it requires high enough viscosity of dough in order for the shear rate and shear stress within the extruder to generate thermal energy. The following formula may explain the dough behavior under extrusion conditions with respect to development of thermal energy.

$$\eta_{app} = \tau/\gamma$$

where η_{app} is the apparent viscosity of the dough within the extruder, τ the shear stress, generated by pressure within the extruder, and γ is the shear rate. The shear stress and shear rate are further explained as:

$$\gamma = V/D \quad \text{and} \quad \tau = P/A$$

In the above formula, shear rate, γ is defined as V, i.e. the velocity of the plates adjacent to each other which is equal to the velocity of the shoulder of the screw tips against the wall of the barrel of the extruder, divided by D, i.e. the distance between the plates or distance between shoulder of the screw and the barrel wall. Shear stress, τ, is defined as P, i.e. the pressure between the plates, over the area A, where pressure is being induced. Each component of the recipe ingredient has a certain effect on the final dough viscosity within the extruder. The net result is varied based on the concentration of each component such as sugars, proteins, fats, carbohydrates and fibers as shown in Fig. 9.1, as well as the pH of the dough within the extrusion system.

Figure 9.1 shows the effect of the addition of a. 20%, b. 40%, c. 60% soya isolate to soya flour while the processing conditions are kept constant and the magnifications of the micrograph are the same at 200×. Micrograph d. shows the control or extruded soya flour with no additives. Micrographs e, f, g, show the effect of the addition of sugar in the form of sucrose to soya flour in 5%, 10%, and 15%.

The above micrographs show that the addition of proteins, such as soya isolate, tends to increase the dough viscosity within the extruder thus resulting in more heat generation within the extruder and melting of the pertinacious globulins to form a fibrous-like matrix which is indicative of protein texturization. On the other hand, the addition of sugars to the dough tends to reduce the viscosity of the dough within the extruder thus resulting in less heat generation and less melting of proteins and starches and a weaker matrix. The higher protein matrix requires a greater force of deformation for a hydrated piece. In dried foods a much higher crunch can be realized with products containing higher protein content. In hydrated texturized protein foods a chewier final product can be realized and is usually marketed today as meat substitute.

Under the above scenario, if a starchy-based product were extruded under severe conditions the viscosity of the final product would be lowered due to over shearing or grinding of the granules. This phenomenon can be demonstrated very easily by the extrusion of a given product and measuring its viscosity after each extrusion run. This test will indicate that the viscosity of such products tends to drop considerably each time the product is passed through the extrusion system. The extrusion temperature at the die will also be reduced for each consecutive extrusion run due to lowering of extrudate viscosity.

In order to show the effects of varying pH of the raw material on the textural profile of the final product we conducted the following tests. The raw soya flour was hydrated to 30% moisture by the use of water at pH values of 4, 6.5 and 9.0. The pH of water addition was adjusted by the addition of sodium hydroxide and hydrogen chloride. The raw hydrated soya flour was checked for pH by the dilution of one part hydrated soya flour to nine parts of distilled water. After extrusion the product was tested using the Ottawa textural cell. The results were a number of picks and areas where the force imposed on the product was tested.

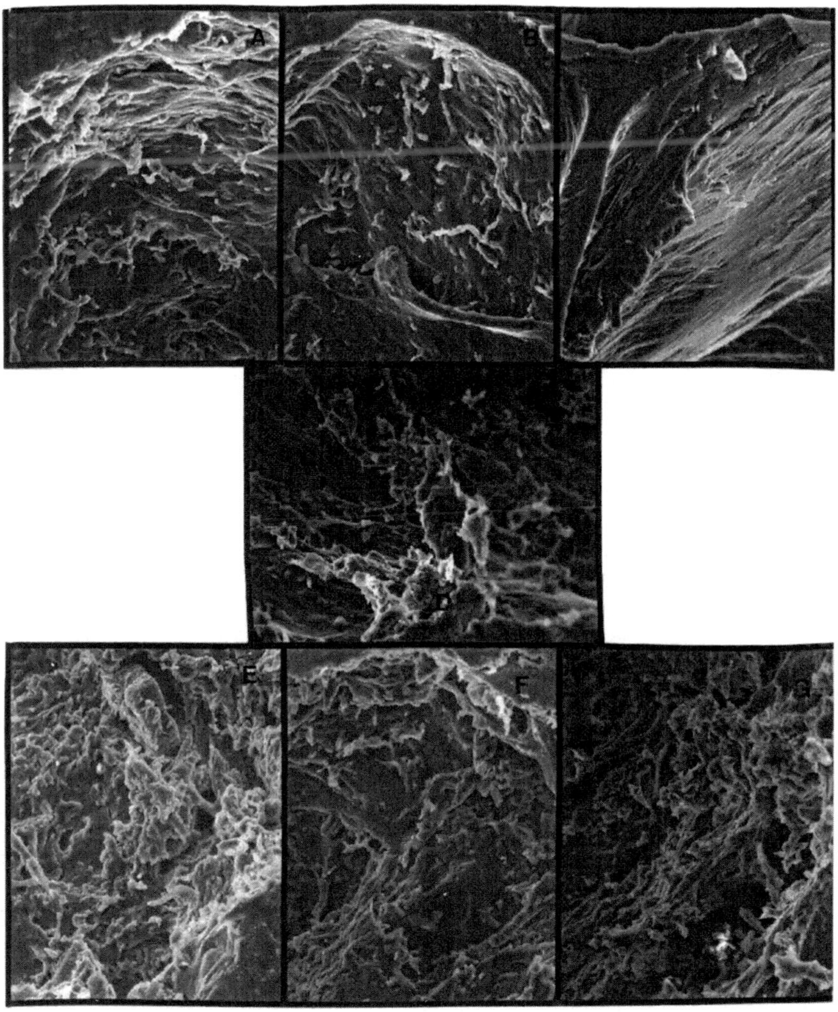

Fig. 9.1

The results are shown in Fig. 9.2. This figure is a three-dimensional graph of the stress values of extruded soya flour at the three different pH levels (4.5, 6.5, and 9.0) as a function of the extruder barrel temperature, ranging from 82–87ºC, no heat added (NHA) to 154ºC. Stress is measured in N/cm^2, and is a measure of the maximum force required to push the extrudate through cell grids (N)/the plunger area (cm^2).

One conclusion which can be gathered from the above results is that the pH of the raw material is essential for the viscosity of the extrudate within the extrusion system, thus resulting in a better melt at higher pH values and more textural development at higher temperatures of the extrusion system. Certain

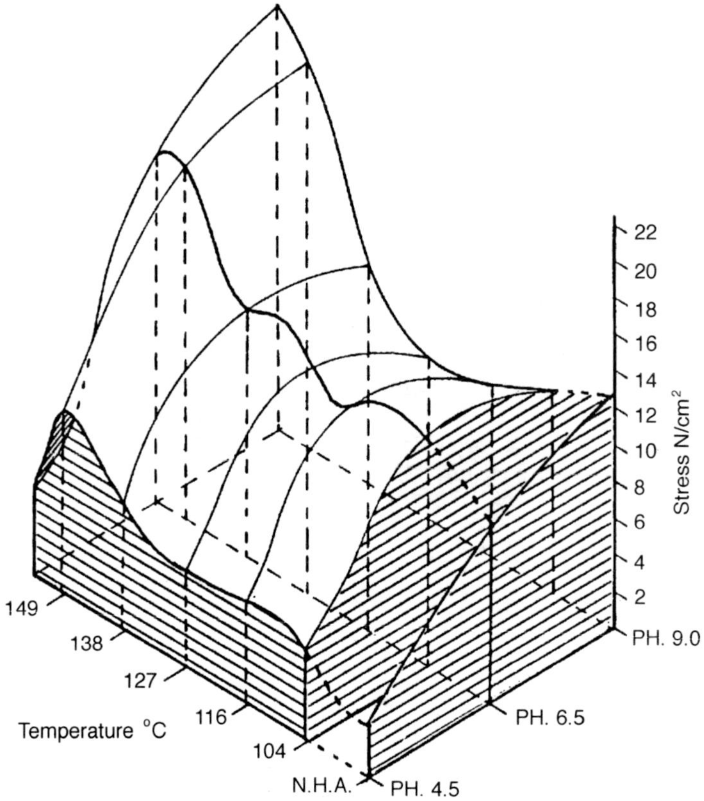

Fig. 9.2

dips can be seen in the overall three-dimensional image which indicates that the process is not linear but that at various pH levels the proteins and carbohydrates will have different viscosity and solubility. This is reflected in the dips that are shown in the lower extrusion temperatures. Under the NHA condition there is a relatively linear increase in the textural profile as the pH of the raw material increases.

Based on the various studies conducted at numerous universities and the need for a mathematical modeling required to explain the process of single-screw cooking extrusion, three sources of energy can be isolated to be responsible for the generation of heat energy necessary for the cooking and gelatinization of a starch grain product during extrusion.

Conductive energy

Conductive heat is defined as the heat energy that is conducted to the extrudate via the jacket. It is irrespective of how this energy is generated. Also minor conduction can take place from the screw core to the product, as the screw gets hot at one point of the extruder due to convertive energy and transmits that to the rest of the extruder. Due to the short residence time of product within an extrusion system, the energy input from conduction is very limited. The use of intermeshing, co-rotating, self-wiping twin-screw extruders can influence the extent and role of conduction energy; however, it is still limited to approximately 90–100 Btu/hr. ft, depending on the extruder size and type.

Convective energy

This is the portion of the energy that is put into the product by means of direct steam injection, hot water, hot liquid, or gas injection either at the conditioning point or directly into the extruder barrel. Such an approach increases the efficiency of the input of thermal energy into the product within the extruder, thus not only increasing the extruder capacity, but also providing direct thermal energy to hydrated grain particles to gelatinize, thereby making the starch more elastic and less susceptible to shear damage. Even though such an approach is a good solution to most extrusion problems, it is limited by how much steam can be injected into the system due to limited extruder volume. The new concepts in such energy convection may be through a microwave system whereby the portion of the extruder working as a holding chamber is bombarded by microwave to increase the thermal energy of the extrudate.

Conversion energy

This term usually refers to the harvesting of mechanical energy from the conveying and grinding action of the extruder. The process of conveying, pressurizing, mixing, kneading and grinding provided within the extruder and screw profile design with respect to the barrel, or in case of twin-screw extruder with respect to the barrel as well as one screw to the other, can generate maximum thermal energy per pound of product. This is proportional to the maximum electrical energy available via the size of the drive motor in the system. This energy is usually measured by the amount of amperage shown on the amp meter. This process generates thermal energy through shear stress and shear rate based on the apparent viscosity expressed in the following power law model for viscosity:

$$\eta_{app} = \tau/\gamma \quad \tau = m\gamma^n$$

where η_{app} is the apparent viscosity of the dough within the extruder, τ the shear stress, generated by pressure within the extruder, m the flow consistency, and n value estimated to be n–0.36 and consistency m estimated by

$$m = 0.73 \exp(-14M) \exp(7884/T)$$

where M represents moisture on a dry basis, and T is the absolute temperature of the product. It should be noted that the flow index and the flow consistency values are merely estimates and that, during the extrusion process, product goes from flour consistency to a high, viscous, dough-like product as the raw material is mixed, cooked, and sheared. Therefore, it is apparent that not only viscosity behavior changes, but also the flow index is shifted. If the starch granules and the protein globules are not well hydrated and are not elastic enough to resist the shear rate during extrusion, the convertive energy will cause great damage to the starch granules and protein globules and will result in a product which is rapidly water soluble and popcorn-like in texture for starch-based material and metallic taste for the high protein-based material. Damaged starch will result from a poor grinding system or a poor extrusion system. The final products will exhibit characteristics of rapid water absorption, gaminess in the mouth, and very short bowl-life for cereals.

To understand better the effects of conversion energy, it is important to evaluate how the energy is absorbed by the product. Convertive energy can be absorbed by two distinct methods within the extrudate: first, by means of the shear stress generated in pressurization chambers, and second by the shear rate which is generated by the rate of screw revolution and chambers where the screw and barrel or screw and screw are in close vicinity to each other. Both generate heat energy. While shear rate is destructive to the starch and protein particulate, the shear stress is more desirable because a pressurized system assists in the forceful hydration of the granules. The conversion energy is absorbed by the product in three distinct pathways:

1. Pressurization of the product results from conveying or pumping work done by the screws. Highly proteinaceous products usually require higher pressures of 1,500 psi in order for the moisture and thermal energy to infiltrate totally into the protein globules and to allow the protein matrix to be formed and elongated into fibrous texture at the die. While for starch granules very low levels of pressure are required in order to achieve expansion, achievement of a specific point of pressurization for plasticized product is essential in order to utilize the high pressure steam bubbles within the continuous medium to infiltrate totally and affect all starch molecules within the matrix. Otherwise, variations in the extent of cooking may occur where the outside of the pellet may be well sheared to form a continuous texture due to addition of shear at the die wall, while the inner matrix of the pellet may be under-sheared and under-cooked, thus forming a harder inner core and a skin on the extrudate during exiting from the die.
2. Sensible heat is the energy used to increase the temperature of the product within the extruder. This reaction can be determined easily by placing a thermocouple to detect the product temperature in isolation from the barrel and the environment. This energy source may change due to the ingredient make up and the boiling point of the liquid plasticizer within the ingredient formula.
3. The heat of reaction is the energy required to drive a chemical reaction. Since cooking is an endothermic reaction, it refers to the amount of energy absorbed by the product components to proceed with the reaction. This part of the heat absorption has been described by Harper and Holoy (1979) in the following equation:

$$E_t/m = \int_{T_1}^{T_2} C_p dt + \int_{p_1}^{P_2} dp/\sigma + \Delta H^0 + \Delta H_{st}$$

where E_t/m is the total energy (E_t) absorbed per unit of mass (m); T is the product temperature; C_p is the specific heat of the product; p is the pressure; σ is the density of the product at pressure p; ΔH^0 is the heat absorbed by cooking reaction; ΔH_{st} is the heat of fusion of lipids; and subscripts 1 and 2 refer to the start and end of the process. More simply put by McCabe and Smith (1956),

$$E_t/m = \int C_p\, dt + \Delta H^0$$

In cereal cooking, the main reaction is starch gelatinization and protein denaturation. The endothermic heat of reaction has been reported to be 90–100 kJ/kg for proteins (Harper, 1981) and 10–19 kJ/kg for cereal starches (Stevens and Elton, 1971). According to Caldwell and Fast (1990), if no real data is available for heat of reaction, then the total heat of reaction can be estimated from the average heat of reactions.

$$H = 14X_s + 95X_p$$

where X_s = starch fraction and X_p = protein fraction of the formula. Specific heat of a product is a function of temperature as well as composition. This becomes significant in a continuous extrusion cooker where high temperatures are often reached without real need. Shepherd and Bhardwaj (1986) present a more accurate formula based on experimental data for cereal grains as follows:

$$C_p \approx 1.424X_c + 1.549X_p + 1.675X_f + 0.837X_a + 4.187X_m$$

where C_p = specific heat of the product kJ/(kg·K) and X = weight of fractions of formula designated (X_c = carbohydrate c; X_p = protein p; X_f = fat f; X_a = ash a; X_m = moisute m) by further simplification to yield the following:

$$C_p \approx 1.71[X_m(4.187 \cdot C_d)] + 0.00304T - 0.292$$

where T = temperature in °C; C_d = specific heat of the dry matter (kJ/kg K); and X = weight fraction of constituents.

It is obvious from the flow and dynamic values of the extrudate viscosity that all three sources of energy input are not interchangeable. They contribute to the total energy balance in a different manner. Mechanical energy and conductive heat energy are both transferred to the product by the surfaces of cooking extruder barrels and are associated by the heat transfer coefficient, which in turn, is influenced by wiping and resurface generation of the product at the heat exchange point of the inner barrel. From reported results it is apparent that starch gelatinization can occur at temperatures of 180–200°F and that air cells can be formed at the die temperatures of 220–230°F. If the additional heat energy provided for the product medium is overdone, it can cause earlier plasticization of the starchy product, thus providing an increase in the shear stress levels of the system.

Twin-screw extrusion

Under twin-screw extrusion the mathematical modeling tends to shift more towards the conductive energy if the screws are designed correctly and the extruder is operated under lower rpm. This is true due to the fact that there is more surface area in a twin-screw extruder than in a single-screw extruder as well as its mechanism in propelling the product forward. Within a twin-screw extruder the product pathway can be designed such that the pumping action can be utilized best after the plasticization of the extrudate thus imposing higher pressures in the subsequent stages of extrusion. We can then apply the shear stress portion of the mathematical model for extrusion. This will introduce a condition that will mimic the baking parameters and, as this effect is magnified and prolonged, more of a baked taste and conditions within the dough can be realized in the final product. Due to the most advanced designs in twin-screw extruders they tend to be very high torque machines with maximum ease of operation and a true self-wiping process. The extruders are so versatile that an infinite formulation of screw design and processing parameters can be achieved. The versatility of such a system makes this process well accepted in various food processing markets from dutching of cocoa to pre-gelling of cereal flours, and from making bread crumbs to the production of cookie dough.

Very low shear extrusion

Very low shear and long residence time extrusion with US patent number 5,914,148, also known as no shear extrusion is one of the new concepts which has been developed and patented. This process utilizes the twin-screw extrusion system for pumping action to move and plasticize extrudate to the next chamber which then uses high conductive energy to heat the product as it moves forward in the cooking chamber without much shear. This plug flow within the heating chamber can have a residence time of 8–20 minutes at pressures of 500–1300 psi. The product is then forced through a number of holes or shapes to produce a fully gelatinized extrudate which can then be puffed as a ready-to-eat product or flaked and dried and ground for pre-gelled markets including baby foods. The viscosity and values received from Rapid Visco Analyzers (RVA) have indicated a very high viscosity profile obtained using this technique of extrusion. A further benefit from such a system is the flavor development during this long conductive heat condition. The products using such a method produce a very long bowl-life and a very crunchy end product. The development of this processing technique which utilizes the long residence time and low shear has opened new possibilities in to the extrusion process. The new area of application is flavor development and debacterialization of the seasonings and herbal flora. High bacterial count, as well as other types of pathogens common in organic farming, become apparent within the herbal and other raw materials making the use of this process most applicable for sterilization.

9.4 The market for baby foods

Baby foods and cereals can utilize a number of the above processing techniques in production. However, it is important to first establish the market available to the consumer in baby foods and cereals and define the characteristics of that market. We can easily separate these markets into two categories:

1. baby cereals
2. baby foods.

Highly soluble baby cereals are commonly extruded with high shear rate and formulated to use fruit juices, cane sugar, or fructose as a sweetener and fortified with stable vitamins and minerals. Such cereals are formed into specific shapes and designs and use little or no preservatives or coloring. Most are in the organic and natural section of the store and may use natural preservatives such as vitamin E, rosemary and the like. They are usually marketed in a bag-in-a-box and consumed with dairy milk or soya milk that are formulated for toddlers as well as younger children. The product may be direct-expanded or flaked. It may also contain pieces of dried fruit or contain high calcium phosphate, with a variety of natural sweeteners sprayed on the surface of the product. The product needs to have a very short bowl-life and very short crunch. The morphological and ultrastructural view of such a product will reveal very minute air cells in the cross-section with thin air cell walls. The extrudate will absorb the liquid milk readily and will turn to a spongy soft product holding its shape when extruded. The textural profile of such a product is essential to the acceptability of the consumer.

Extruded baby foods are highly soluble fully gelatinized products with various natural sweeteners added to the mix to produce a fully nutritious and fortified product used as baby food supplement. The product is usually not in any form or shape and may be flaked and ground to a semi-powder form. The fully ground gelled product is easily hydrated and can be mixed with hot, warm, or cold water to form an oatmeal consistency to be consumed within a short time. The most popular of these products are commonly found in the organic and natural food stores as well as regular supermarkets. These cereals, after grinding, can be mixed with a number of other ingredients to form a complete and highly fortified meal for babies and toddlers. The major advantage of such a mixture is that the extruded portion of the product is sterile, assuming a clean grinding system. When grinding the extrudate it is important to limit very low or no cross-contamination via handling or transporting. However if other additives are introduced after production of the ground cereal the possibility of foreign materials entering into the product with pathogenic implications is much higher. A good precaution is to extrude the complete mix, thus allowing the pathogens to be destroyed during the high temperatures and pressures of the extrusion system, leading to a more secured clean product.

In both of the above baby food systems, a liquid is added to the product, allowing the food to hydrate before it is served. It is important to insure and educate the consumer that any added liquid, whether it is milk or water, must be clean and pathogen free. After the addition of such liquids there may be a delay in the consumption of the final product which can have catastrophic results if precautions are not taken to avoid high risk conditions. This may not be much of an issue in developed countries. However, developing countries, especially areas where refrigeration systems are scarce or the drinking water can be easily contaminated, may be a major problem. The liquid can easily be considered unfit if it is slightly contaminated and mixed with a highly processed cereal containing some sweeteners at room temperature. Such a system in liquid form at room temperature can result in the rapid growth of pathogens. This will result in an unfit product within a very short time.

9.5 Baby food products

Other products that are relatively new to the toddler or baby food market are snacks designed to be held easily by the child, such as a three-inch stick. These sticks melt and become gummy once in contact with the child's saliva. These products are usually cookie type recipes with similar texture and taste. Cheesy pretzels are common as well. They may be formed into long sticks of 3–4 inches, which are filled with jam. The contrasting texture between the pretzel on the outside and the jam-like filling on the inside adds textural appeal of the product for the child. These fillings may be fruit-based with sweet and sour flavor. The outer cereal casing of such a snack is totally gelatinized and formed into a soluble cereal which may smooth the transition of the sourly sweet taste to the child's mouth.

Products made up of mixtures of fruit purée and fruit juice, along with other ingredients such as brown sugar, cane sugar, and wheat flour cooked together can provide an interesting final product that is wholesome and a good source of energy to toddlers. These snacks can be made into a three to four inch thin stick form that dissolves slowly as the toddler sucks on them. Outer coatings of such products are usually dusted by various flavorings, sugar powder, and starch. Because they are a high moisture product of 18% or so and contain high amounts of sugars and sweeteners, they may not be stable if left out for a long period of time. They need to be packaged in a manner where only a small portion of the product is removed from the package and the package is then resealed to prevent hydration of the sugars from atmospheric humidity.

The sizes and shapes of these foods are also of great importance due to the limited size of the mouth and throat of the young consumer. When the shapes or the sizes of these products are being considered, the possibility for the child to choke or to suffocate on the finished food must be borne in mind. Such designs should be used which would allow easy usage and safe consumption once the child has the product in his or her mouth. A good example is the use of holes in the middle of the food pieces, such as O-rings of 10–25 mm or sticks of 2.5–8.0 cm in length having a hole in the middle.

The recent development of cereal with milk protected by US patent number 5,894,027, is another advancement in the field of baby foods, which should be considered. Such products can utilize a pre-mixture of fully gelatinized, cooked and formed cereals combined with surface coating of cold water soluble powder which can dissolve readily by adding water into the bowl giving a milk-like flavor and mouthfeel when consumed by the child or grown up. The cereal that is coated with the milk substitute can be formulated or produced so that it will hydrate and partially dissolve readily for toddler market or stay crunchy in the milky water thus having a long bowl-life designed for the adult market. By adding the required vitamins and minerals as well as adjusting the protein profile in the recipe, and topical application, such cereals containing milk look-alikes can bring to the consumer in any society and region of the world, young and old, rich and poor, a range of benefits that can satisfy hunger and deliver needed, well-balanced nutrition. Due to the ability of such products to be made from large sources of raw materials already abundant in many parts of the world, the cost of manufacturing can be minimal.

Syrup infused cereals protected by US patent and European publication number WO98/19562, can produce in one step during the manufacturing process cereals and other foods that are infused with syrup or fruit juice using the twin-screw extruder system. This processing technique eliminates the possibility of cross-contamination that can take place when applying a spray-on coating after the cereal is heated, cooked, and dried. This process produces unique and well-formed pieces of food, with low bulk density, and an extensive crunch already coated and infused with sweeteners. The manufacturing and maintenance of such systems are slightly more complicated but much less extensive than the present coating system that is common to the cereal industry. Use of fat-based coating as well as double or triple coating tanks for various coating materials can be utilized in order to give a unique texture to the end product.

An under-oil cutting technique is another unique textured product that is produced using the US patent application 5,527,553 issued on June 18, 1996. This process utilizes various extruders such as a single- and/or twin-screw extruder to produce a final product that expands into a high temperature oil-based medium. While the product exiting the die may be at temperatures of 300°F, the oil-based medium may be at 250–365°F. This higher temperature above boiling allows the product to be further dried, infused, and coat the cereal with an oil-based medium. The oil-based medium may contain other ingredients such as proteins and sugars, etc., that will be incorporated into the finished product. The final result of such a process is a product that is dried and totally sealed from a water-based environment, thus allowing the product to be used as inclusion into other high moisture-based foods such as ice cream. One such application is the use of a fat-based yogurt coating, or chocolate/white chocolate compounds in the high temperature liquid-medium. In this format, the steaming moisture from the drying effect of the extrudate is easily separated from the transporting liquid medium at the die area. Using this technique the extrudate, as it exits the die, is immediately introduced to a high temperature water immiscible liquid medium. This liquid medium will carry the product away from the die to a final destination, while continuing to remove moisture from the cereal due to high temperature surroundings, and finally separating the cereal from the liquid medium via various means of separation technique such as sifting or spinning. The moisture driven off the cereal is then separated from the hot medium via a vent system thus achieving a drying and coating method both in one step without being handled by the extruder operator. The product is then cooled and packaged.

9.6 Processing benefits of twin-screw extrusion

Process control is becoming the single most important issue in any extrusion system. This is because the extrusion system is dynamic and has a short residence time, meaning that accurate control (real-time control) is essential for the steady production of high-quality finished product. The twin-screw extruder is probably the most advanced and widely used system that takes advantage of such an accurate newly developed control system. To achieve the precise control on such systems the use of a fuzzy logic controller becomes essential to this process as well as extremely sensitive and real-time sensors. Under such a scenario, the process is self-adjusted and controlled for repeated start-ups and shutdown and during 24-hour continuous production based on empirical values. A layman with a small investment can enter the arena of food processing and food manufacturing using this controlling procedure. Although there are number of fields of education, such as process engineering food technology and cereal science, needed to utilize the system fully for best processing conditions, nevertheless the lay investor can easily compete at a small level with large corporations on manufacturing various cereals and food products by gaining some education for his technicians from the equipment manufacturer.

The major benefits of using the twin-screw extrusion system coupled with a good systems control program, is the continuous nature and ease of operation. Specifically, if we keep the raw ingredients consistent both in production and processing, the final product should not deviate from the targeted product during processing day in and day out. In the manufacturing of baby foods, such a system in a developing country can be a life saver due to its continuous nature, its ease of operation, and its self-controlled parameters, as well as closed systems engineering which prevents the product from coming into contact with operators at any time, thus protecting against cross-contamination. The twin-screw extrusion system can be an extremely closed-loop processing system with wide versatility. It does not require a large processing area due to its small footprint. It is highly efficient in converting electrical and thermal energy to a cooked product. Finally, it is the least labor intensive among all the past processes in the food industry. It lends itself easily to research, development and scale. Profiles of most products are easily achieved with minimal work. In most cases, the end product is shelf stable for at least six months and can be packaged with a minimum packaging requirement. The consistency of producing a fully gelatinized, highly sheared and dexternized final product on a continuous and consistent basis makes this process an ideal equipment of choice for AID and World Bank types of programs designed to improve the quality of food and nutrition around the world.

9.7 Socio-economical future of baby food production

One of the major dependencies of developing countries is the source of food for their young population. In the USA as well as Europe, extensive fund raising has been going on from the private sector to buy food in order to feed the young population of the developing world. These are certainly meritorious, well-intentioned projects. However, the need is growing and to date these projects have not proven an efficient means to lessen the dependence of the populations of developing countries on handouts. It is from this point of view that in this chapter we would like to emphasize the process by which to make various societies and populations independent of foreign aid. Utilizing local grains and eating habits to produce nutritious finished products that are acceptable by the locals can extend the shelf-life of a grain from a couple of weeks to six to nine months. This has tremendous implications on local economy, bringing into demand many local products that can be used as the source of raw materials. The versatility of the twin-screw extruder combined with new developments in its control system and food technology, as well as advancements in extrusion processing techniques, can make a processing and manufacturing company in any socio-economical setting a profitable and workable project. In the European and US markets, a box of cereal may have cost associated with its raw material which is less than the cost of the printing on the box. This makes the value of efficient processing techniques enormous in areas where the raw material is in abundance and requires some consistency in production of raw ingredients in order to fit the needs of such a manufacturing technique. A pre-purchased alignment between the grower, processor, and manufacturer can assure the profitability of such systems and insure the education and independence of such societies around the world.

With some educational training of operators and technicians along with organized and cooperative efforts between the grower, processor, and manufacturer of baby and toddler foods, the developing world can become self sufficient in a short period of time and provide a steady nutritious finished product which would fit the taste profile of the local consumers successfully. The expression, give a man a fish and you feed him for a day; teach a man to fish and you feed him for life, applies here. In the long run it is proven that it is much more efficient to import education and know-how than raw or finished goods into a deprived segment of world population. Under the first scenario the education can improve and advance the condition of the user while the second scenario will bring dependency and larger problems for the future to solve. While all other advancements and exports of goods may be seasonal or limited in duration, food processing knowledge and manufacturing is ever enduring and has lasting implications through time. Within the food production market, the key lies in the processing and manufacturing of the finished food product due to its ability to extend the shelf-life and improve the digestibility and safety of food. It is the basic fuel for the human machinery. It is essential to make one profitable in one's own society. Lack of such fuel or low-quality fuel can contribute to the advent of a lethargic society leading to degradation of human living quality in a short time. Ownership of the knowledge of food production and processing is the single most important incentive for a society to advance and improve its other conditions.

9.8 Conclusion

Baby and toddler foods are in a fast growth segment of the market in developed countries. These growth areas are realized in most easily the specialty organic and natural food areas. The processing techniques for the production of such products which require high viscosity when milk or water is added is commonly known to use twin-screw extrusion and various other techniques developed for such systems. With today's advancements in the area of systems control and the new approaches to food processing, the entrance of a layman with some financial means is becoming easier and more realistic. This new advancement in the area of food processing has opened a new area of responsibility for the advanced countries as well as financial and educational institutions. The dilemma of whether the raw and finished goods or the knowledge and means of self-independence be used as a helping technique to the developing world is being answered by the new advancements in technology. Responsibilities are there for world lending institutions as well as world assistance to bring such systems of twin-screw extruders into developing countries and allow them to educate themselves in the production of baby and toddler food. By allowing these countries to enter and manufacture their own processed foods we are opening new dimension of progress to humankind, which at the end will be beneficial to the whole population of the Earth.

9.9 References

CALDWELL, E. F. and FAST, R. B. (1990) *Breakfast Cereals and How They Are Made*, St Paul: American Association of Cereals Chemists Inc.

HARPER, J. M. (1981) *Extrusion of Foods*, Boca Raton: CRC Press.

MCCABE, W. L. and SMITH, J. C. (1956) *Unit Operation of Chemical Engineering*, New York: McGraw-Hill.

SHEPHERD, H. and BHARDWAJ, R. K. (1986) Thermal properties of pigeon pea, *Cereal Foods World*, 31: 466.

STEVENS, D. J. and Elton, G. A. H. (1971) Thermal properties of the starch/water system. I. Measurement of heat of gelatinization by differential scanning calorimetry, *Statch/Strärke*, 23(1): 8.

Index